D1741531

# Mathematics through School

The cover design is based on one of the Optical Coincidence Feature Card Systems marketed by George Anson and Co. Ltd.

## Contributors

### Members of the Shell Mathematics Unit at the Centre for Science Education

Margaret Brown
Brian Dudley
Bob Lewis
Tony Malpas
Geoffrey Matthews
Alaric Millington

### Other contributors

Barry Blakeley
Highgate School

Julia Comber
Headmistress, Thorntree Infants' School, Charlton

Jan Evans
Lavender Hill Girls' School

Margaret Hayman
Putney High School for Girls

Sheila Morrison
Lavender Hill Girls' School

Wojciech Mrozowski
Wandsworth School

Stuart Parsonson
Harrow School

Elizabeth Rowland
St Catherine's Middle School, Merton

David Taylor
Assistant Director, School Mathematics Project

# Mathematics through School

Edited by **Geoffrey Matthews**

**John Murray** Albemarle Street London

© M. Brown, G. Matthews, T. A. Millington, 1972

All rights reserved. No part of this publication may be reproduced, stored in a retrieval system, or transmitted, in any form or by any means, electronic, mechanical, photocopying, recording or otherwise, without the prior permission of John Murray (Publishers) Ltd., 50 Albemarle Street, London, W1X 4BD

Text set in 10/11 pt. Monotype Times New Roman, printed by letterpress, and bound in Great Britain at The Pitman Press, Bath

0 7195 2700 7 Cased
0 7195 2701 5 Paperback

# Preface

Something dramatic has been happening to school mathematics, and 'revolution' is perhaps not too strong a word. But revolutions generate fog, and the New Maths is hard to pin down: it has been taken to mean anything from a catalogue of unfamiliar symbols to the vague notion of 'letting the children think'.

This book is designed as a modest exercise in communication. It aims not only to give that long-suffering character, 'the intelligent layman', a glimpse of the new spirit, but also to show the embattled teachers themselves something of what's going on at other levels. It is not possible here to give more than the flavour, and the interested reader must be referred to the publications of the new projects themselves. (See *Notes* at the end of this book.)

The material is based on a series of lectures given at the Centre for Science Education during the winter of 1970/1. These have been edited into a form more suitable for print, but no attempt has been made to achieve a consistency of style throughout, as each speaker was making a personal statement about his experience. The chapter headings may seem to perpetuate the arbitrary divisions in our educational system, but their familiarity will at least help the reader to identify a particular section which he wishes to explore.

I express my appreciation to Bill Sheridan, of the University of Reading School of Education, who helped with the thankless task of transforming the lectures into print.

G.M.

# Contents

# Plates

*All the photographs were taken by Julia Comber*

# 1  Pre-School and Infants

Geoffrey Matthews
Julia Comber

The infants' teachers were the first to recognise the truth of two very important propositions:

> All children are different.
> Young children learn by doing.

The first of these inevitably points to the idea that children should be working in small groups or individually for at least some of the time. In many infants' schools this has led to the adoption of 'integrated programmes', from which the rigid timetable has been removed and very few fixed times remain, such as use of the hall for P.E. and assemblies. For the rest of the time painting, reading, modelling, writing and mathematics go on alongside each other, meeting the requirements and interests of both children and teacher. If the teacher feels it is appropriate, it is possible for a child to spend the best part of a day, or even more than one day, on an activity which has reached an exciting stage. The teacher accepts the responsibility to provide a balanced range of activities over a period of time.

The second proposition—'Young children learn by doing'—is beyond question. No child has ever seen the number 3 (nor any grown-up, for that matter); children learn about 3 by handling a large number of trios, three bricks, three toy cars, three bus tickets, three buttons, three beads . . . and from all this activity they abstract the notion of '3'. A good rule in all this activity is

> 'The younger the child, the larger the apparatus.'

The geometry of the nursery is essentially three-dimensional: no small child is very interested in triangles, but he is quite prepared to explore a tetrahedron if you call it a tent and let him get inside it.

If the building bricks which you thought might make a good castle turn into a car instead, so much the better: the children are acquiring a feeling for space.

We must now ask: what is mathematics? Is it, for example, this?

$$27$$
$$46\times$$

Certainly this is part of mathematics; whatever 'new maths' may be, children should still be able to do 'sums' such as this one. But the trouble in the past has been that such topics have been introduced in the wrong order and at the wrong rate. An important discovery has been *how much there is to learn* before numbers can be handled with understanding, and how by apparently starting more slowly children will have got much further than before, with greater ability and confidence. One message for us all is this: before complaining that children haven't 'done' a topic, we ought to ask, have they grasped the relevant concepts?

Let's come back to our 'sum'

$$27$$
$$46\times$$

To understand the calculation of the 'sum' it is necessary to appreciate that $27 \times 46$ is the sum of $27 \times 40$ and $27 \times 6$. But then what about '27' itself? It really is not obvious to a small child that this means 'two tens and one seven': there is no logical reason why it shouldn't mean 'two and seven' or why it shouldn't be the same as '72'. The difficulty of this idea of place-value has been greatly underestimated, even though an adult impatient with his small child is quite prepared to be indignantly mystified when told that a computer would recognise '10' as 'one *two* and no units', i.e. 2, rather than 'one *ten* and no units', because the poor thing can only be *on* or *off* and doesn't have ten fingers. Even before '27' the child must understand '2' and '7', must have abstracted the concepts of 2 and of 7, the 'twoness of 2', the 'sevenness of 7'. One of the basic activities leading to this understanding is 'one to one correspondence' or matching: 2 drivers are matched with 2 cars, 7 cups with 7 saucers and so on.

So *matching* is a vital experience in an infants' classroom. It is one of a network of early mathematical concepts shown in the map on pages 4 and 5. This is based not only on forty years' work of Piaget and his colleagues at the Institut des Sciences de l'Education at Geneva but also on much experimental work by the pilot areas of the Nuffield Mathematics Project (age-range 5 to 13).

One significant feature of this map is the amount of 'spatial' ideas: neighbourhood, distance and length, angles and so on. Perhaps in the past we have concentrated too exclusively on the sacred idea of 'infant number'—a strange expression, if you come to think of it. All the same the concepts leading towards number are very important: for example, *ordering* leading to the sequence 1st, 2nd, 3rd, and as we've seen, *matching* leading to 1, 2, 3, 4. . . . Perhaps the best way to illustrate some of these concepts and how infants' teachers provide experiences leading to them, is to give some snapshots showing them in action in the classroom. Similar examples will arise over and over again until the particular concept is fully acquired.

Sorting is a fundamental activity which can be started very early: for example, 5 year-olds sorting out the dolls, the cars and the other toys. A little later, a miscellaneous collection brought in from a visit to the local park can be sorted in many ways which can be discussed with the teacher (*Plate* 2). There are many opportunities for more complicated sorting: for example, at Christmas the children taking the part of angels in a nativity play sorted their attributes and charted them according to length of hair and colour of eyes. Sorting leads naturally on to *inclusion* and again at Christmas some children sorted the gifts they had made so that the car was included in the set of toys which in turn was included in the whole set of 'Christmas' things (*Plate* 3). This activity was certainly within the range of interest of the children, but at the same time leading for example to the notion that a set of four objects can be included in a set of seven objects and eventually to the abstraction $4 < 7$, 'four is less than seven'.

Making comparisons is another basic activity. When children are playing with bricks and change from 'that tower is big' to 'this tower is taller than that one' they are on the way to *ordering*. A simple example of ordering would be making models of the three bears of the Goldilocks story and putting them in order of height. More elaborate ordering took place in one class, again at Christmas, when they decided to make a 'Christmas suit' which would be large enough so that everyone could wear it in turn. This involved a good deal of measurement including marking off waist sizes with strips of paper and then putting these in order of length.

One of the many ideas leading up to multiplication tables is *intersection:* for example, an 8 belongs where the row labelled 2 meets the column labelled 4. The idea of belonging to more than one set at once is important in many other connections, for example, a square belongs *both* to the set of rhombi *and* to the set of rectangles. Some children were given an experience of intersection in their

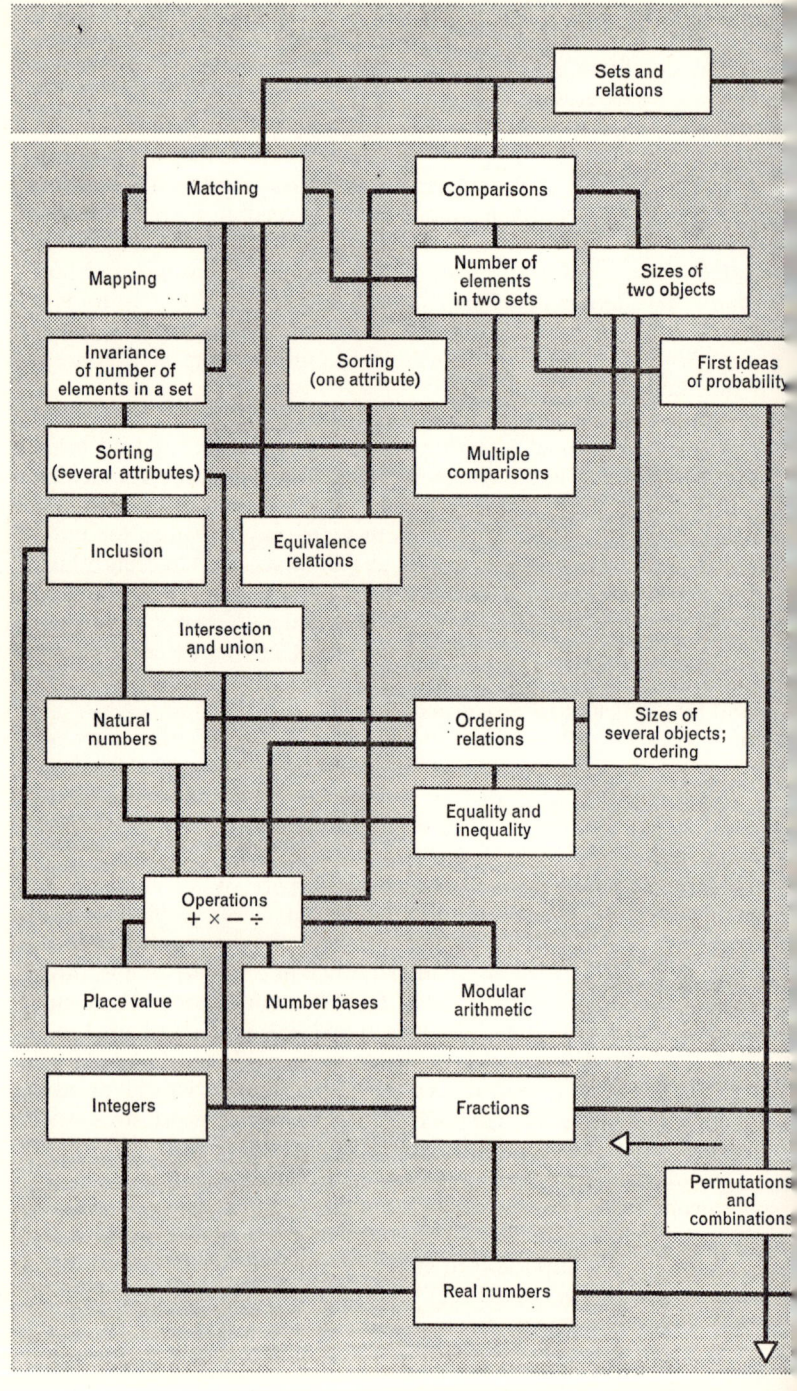

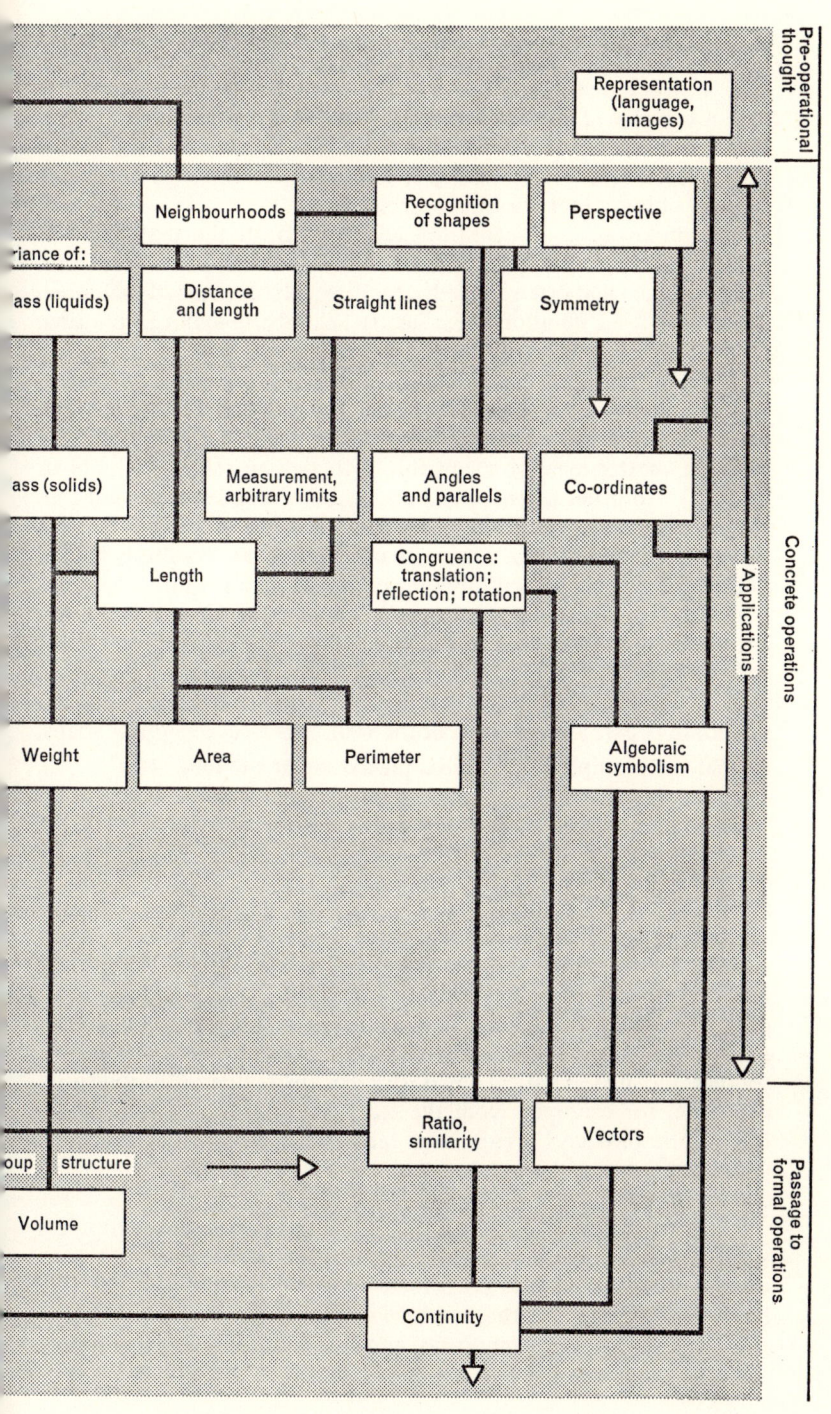

Guy Fawkes pictures (*Plate* 4). Their teacher had made a table with a top row of colours and a left-hand column showing drawings of various fireworks. The children then put their cut-outs in the appropriate square; for example, the yellow rocket was put in the 'yellow' column and the 'rocket' row.

The importance of acquiring concepts, from the mathematical point of view, is largely that they enable computation and measurement to be carried out with understanding. The concept of matching, which has been mentioned earlier, is behind the skill of counting. If a child is counting different-coloured beads (say in an 'Indian necklace' which he has made), he must have some sort of system and be able to match the beads against the numbers 1, 2, 3, . . . (any child can be taught to recite the numbers parrot-fashion, but it is a different matter synchronising this with ticking off, say, beads or buttons). Addition of two numbers, again, depends on the concept of the 'union of two disjoint sets', which in perhaps plainer English means that the children learn about addition by counting up the number of objects in two different piles and then counting the total number when these are put together. When the ideas of addition are forming, a good deal of practice is needed, and it is an absurd rumour that 'they don't do any proper sums nowadays'. A typical activity is given by the number strip: the numerals 1, 2, 3, . . . (up to say 20) in order, along a strip of the wall, and a set of shorter strips, so that for example $\overline{1\ 2\ 3}$ can be placed under the long strip

$$
\begin{array}{l}
\overline{1\ 2\ 3\ 4\ 5\ 6\ 7\ 8\ 9\ 10\ .\ .\ .} \\
\overline{\quad\quad 1\ 2\ 3}
\end{array}
$$

and the child can read off that $4 + 3 = 7$.

Children also enjoy finding, for example, combinations of numbers which add up to 10, and the changes can be rung by adding the numbers appearing on say two dice or on the successive throws of one. The art of the teacher is to contrive a number of similar experiences, which nevertheless seem fresh and different to the children, so that the mathematical ideas can be abstracted. Multiplication can arise, for example, from the story of 'Four-and-twenty blackbirds' (one bird lays 4 eggs, two birds lay 8 eggs, . . .) and much computation can come from the study of patterns in a number square (*Plate* 5).

The concept map does not refer by name to 'conservation', a word which has been overworked and misunderstood: it is always necessary to ask what is conserved under what circumstances. We have preferred to use the word *invariance*. If a piece of elastic is stretched, its length clearly is not conserved, but if a pencil is simply moved round the table, then its length is conserved—and it is perhaps surprising that young children don't always 'see' this. Another example of invariance is given by liquids. During 'water play', children will gradually learn that when water is poured from one container into another of a different shape then (provided none is spilt) there is still the same amount. This seems 'obvious' to us, but if you don't believe there's a problem, try this out on some five-year-olds! 'Invariance of weight' also has to be appreciated, and although weight is a comparatively difficult idea, the children can have their first experiences of 'balancing' as soon as they come to school—if not sooner. 'Invariance' is important: for example, the old 'congruency' of geometry is just an investigation of the circumstances in which a figure retains its shape and size—nowadays we talk of translations, reflections and rotations. Children can have early experiences towards these ideas simply by handling shapes and experimenting in placing them together. Invariance of length is also a necessary concept—it's no good measuring if you think the ruler changes size when it is moved about. First experiences with measurement do not involve the use of rulers or any standard measure, but take the form of comparisons with a variety of 'units' of whatever is to hand. The need for a standard measure will eventually become apparent. In one class, some children were planning items to make for a 'holiday shop', including sandals made of cardboard. This led to measurement round their feet on to squared paper and then comparison of area covered was a natural development. On another occasion, taking advantage of a sunny day, children were working in the playground and great interest was shown in the shadows. This led to discussion of early clocks and the discovery that shadows change with time. String was used to measure the length of the shadow of a stick at different times of the day.

We haven't yet mentioned recording. This includes drawing and graphing as well as writing, and these all become more meaningful when directly linked with the children's activities and experiences. A group of children made the best of a wet day by classifying the rain-wear used by the class and recording the information in pictorial form. Children should be encouraged to record their findings in a variety of ways and this leads naturally on to more formal work involving recognition of symbols and computation. A good deal of mathematics can be brought out of a project. One such project, on post-offices, started in a small way when one child was away ill and the class decided to send him some 'get well' letters. A short walk to the post office was fixed and the route discussed; the purchase of stamps involved calculations with money, and on return the general-purpose large box on wheels was converted into a post office van; paper had to be painted and measured in order to cover the sides. A home-made letter box stimulated further writing and, for example, the mathematics of making envelopes.

With all these activities going on, it is quite a problem for the teacher to keep track of where the children are in their thinking and, of course, sometimes the same topic recurs over and over again with progressively greater sophistication. For example, Maria in the nursery may simply be threading and wearing beads; a year or so later, she will be sorting them out systematically (*Plate* 6) and later still be recording her findings, such as the number available of the various colours. Records of progress must be kept, and it is of enormous benefit if infants' and juniors' teachers get together, so that for example a folder is passed on from one school to the other with notes of progress and perhaps half-a-dozen examples of the children's most recent work. This is more eloquent than reams of reports of the variety 'Fair', 'Very fair', 'Trying', 'Still trying', 'Very trying'.

Infants' teachers trying to find out where the children are in their thinking have been greatly helped by the production of 'check-ups' devised at Geneva for the Nuffield Project and linked to the concept map (pp. 4, 5) earlier. These have had to be adapted from the original research accounts and simplified so that they can be used quickly in a classroom with 40 other children. It is most important that these check-ups should be at the service of the class-teacher, to be used only when she feels the need has arisen. It would be quite disastrous if it were felt that every child should be 'tested' on every small topic—examinations have exerted a baleful enough influence already in this country without proliferating and supplementing the '11 +' with a '5+', '6+' and '7+'. So it must be strongly emphasised that the check-ups are strictly *for the comfort of the teacher*. For example, on 'invariance of length', a child can be shown two rods which he will agree are the same length (as far as the eye can judge) and then one is moved to one side, and rotated through a right angle. If the child reckons they are still the 'same length', he is certainly at least well on the way to acquiring the concept. For the concept of 'neighbourhood' or 'being next door to' a child can be shown a model washing-line (shirts, shorts, coat, etc.) and make a copy of it underneath by choosing the correct cut-out 'clothes' from a pile. This is quite simple, but a tougher test is to copy it 'backwards' (*Plate* 7), that is, to reverse the order so that the shirt is on the extreme right instead of the extreme left. A child who can do this confidently is well on the way. To save the teachers' precious time, the children can often make the materials for the check-ups themselves. Finally, let us look at a check-up for one-to-one

Plate 1. Mathematics activity in an infants' school. The following sequence (*Plates* 2 to 8) illustrates some of the concepts discussed in Chapter 1.

Plate 2. Sorting (p. 3)

...s we have made for Christmas

Plate 3
Inclusion
(p. 3)

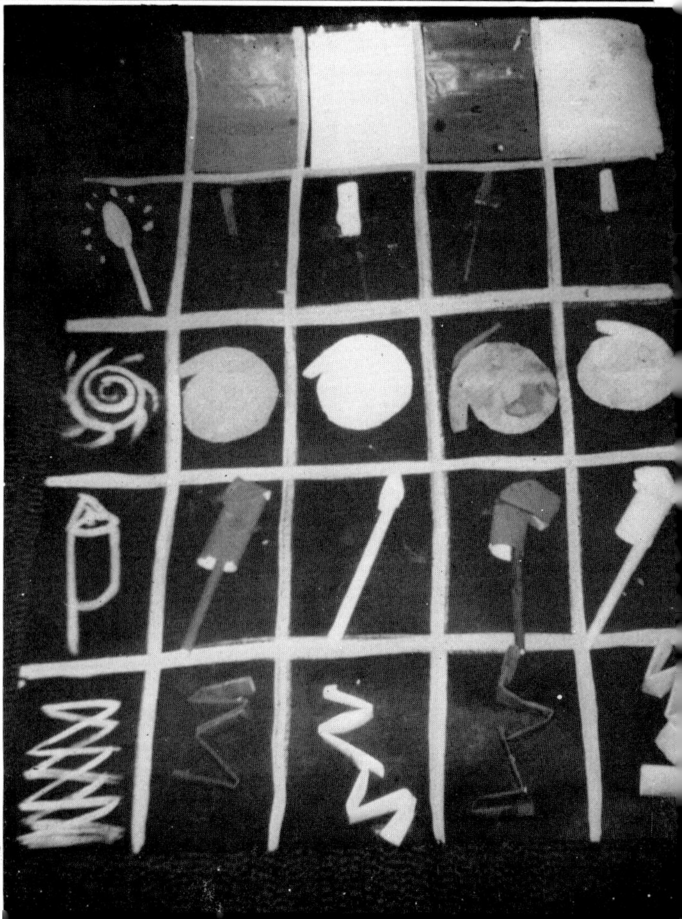

Plate 4
Intersection
(p. 6)

correspondence. The child is asked to match say a penny against each of several items spread out in a line (*Plate* 8). The pennies are then put together in a heap: are there now more pennies or more items or the same number? All three answers will turn up in an infants' class.

We have given an account of some of the mathematical ideas acquired by infants. If a Victorian teacher could somehow come back and visit the classroom she would find it unrecognisable. There is maths with everything: sometimes coming out of a project, sometimes in its own corner, and still, most importantly, with an element of recording, getting down to work and doing sums.

But remember the slogans:

> All children are different.
> Young children learn by doing.

'Doing' leads to 'understanding' and this in turn leads to getting the sums right and, equally important, to enjoying them.

# 2 Juniors

Alaric Millington

The traditional beliefs in all fields of human activity are being challenged so drastically that our role as educators must be closely examined in the light of the changing attitudes in our society. Ideas in such fields as science, technology, sociology and psychology are developing at such a rate that any philosophy of education which is not dynamic, adaptable and free from dogma must inevitably be incapable of developing alongside the great strides being made in other fields. Our concepts of what we now term *mathematics education* have been seriously questioned for a century or more. A hundred years ago, the emphasis on the teaching of techniques was part of the necessary process of educating, if that is the correct term, an illiterate population. It was also associated with the belief that mind and brain were almost synonymous words and the misconception that both could be trained like a muscle. How else could a teacher avoid starvation on a payment-by-results basis of earnings if he did not produce performing monkeys? The 1870 Act which brought educational manna to the masses and the 1902 Act which brought opportunities to the more or less fortunate or unfortunate few, according to which end of the social stratum was your viewpoint, were means to some end and not in themselves revolutions in educational philosophy. In 1931 we began to unbend and some pleaded for a little simple geometry for all children and a reduction in the size of computation, the size being the size of the numbers involved! As recently as 1959, mathematics as a suitable term for use in discussion of primary curricula was first used. Aims began to change rapidly and rote-learning became a dirty word. Throughout the history of state education there have been rumblings of concern about the teaching of different aspects of mathematics and pioneers both inside and outside the system have made gallant and influential steps forward. However, at no time in history have there been more traumatic changes for teacher and child as there have been in the last decade or two.

The developments in scientific and technological fields have created a demand for new mathematics and a reappraisal of the old. The increasing complexities of our society have required mathematical techniques which were unnecessary in the 1930s. The ever-increasing pace of living demands a reassessment of education and its role in human development. We must now accept the fact that most adults will need to change their occupations two or three times in their life and face a longer period of retirement. What this means to society as a whole only the future will tell.

Teachers, especially primary teachers, have been given a precarious and demanding role, because it is largely on their shoulders that the responsibility of developing educational attitudes has been placed. In what form this responsibility will be exercised is anyone's guess but we do know, at this time in history, enough about our changing world, enough about the ways in which children—and adults—learn to be able to assess our role and change ourselves.

What does all this mean to the teacher of our older primary children? In one sense the problems are those o all teachers and a decision to restrict our thoughts to the Britis traditional 7+ to 10+, which incidentally has no relation to a post-Piaget, pre-puberty structure, is an arbitrary one. In many schools there is no recognisable difference between an infant and a junior, though the problem may be physically larger. In some schools they are not sent into orbit at 11 but at some mystical 8 or 9, when they take on the role of middlemen. Much of what we say about mathematics education is applicable to all children, either because it belongs to the field of general principles or because, as was argued in the previous contribution to this series, children are different. There are sound generalisations in the theory and practice of education which are based on considerable research and experience but the wise teacher knows that children learn at different rates and often in different ways. It is, perhaps, better for us to look at mathematics, decide what it is for the primary child and then discuss the kinds of experiences which we, as educators in that vast state between ignorance and wisdom, feel should become part of the climate of primary school mathematics.

One convenient way of viewing mathematics within education is to see it, as A. N. Whitehead did, as the study of relations in number, quantity and space. This does not mean doing sums ad infinitum, converting Imperial units to SI units, or achieving prowess in bisection. It involves considerable experience in all kinds of situations, sometimes selected by the child, sometimes selected by the teacher.

None of these situations is strictly mathematical to the exclusion of other aspects of human experience. Every child, in any situation, is presented with a wide range of sensations, sensory from his external environment, physical, emotional and mental from his internal environment. In this brief survey of mathematics education one can only scan the development of concepts, but as the whole development of mathematical thinking is precisely this, we must look at some of those salient features of the learning process which cannot be ignored by the teacher. Thought cannot occur without concepts and concepts cannot exist without active physical exploration in the early stages of development. I remember some years ago visiting a class of seven-year-olds. The accepted bright child was romping through a sequence of formal divisions by 7 by the simple process of running a finger down the 7-times table. On the other hand the child could not tell me how long each part would be if I broke a 12-inch stick into four equal parts. Another seven-year-old in another school could tell me what a quarter of a 13-inch stick would measure and yet had never heard of the term *divide*. Incidentally, this latter child at a later stage was shown the sequence 1, 2, 4, 8, . . . and he described this as *multiplication by* 2. On being shown the sequence 1, $\frac{1}{2}$, $\frac{1}{4}$, $\frac{1}{8}$, . . . he described this as *dismultiplication*. One child was learning to perform, the other was learning to think. One child was being subjected to a life of cash register mathematics—press the right button and you get the right answer—while the other child was learning through experience of situations which permitted a process of abstraction from concrete structure to mental structure as a prerequisite to solving new problems.

Let us have a look at this process of dividing by 7. Some of us were subjected to the incantation $1 \times 7 = 7, 2 \times 7 = 14, 3 \times 7 = 21, \ldots$ I say some because others never learned this. They learned $7 \times 1 = 7, 7 \times 2 = 14, 7 \times 3 = 21, \ldots$ Some of us even learned the gozintas: 7 gozinta 7 once, 7 gozinta 14 twice, 7 gozinta 21 three times, . . .

A frightening world of notation can result from the use of symbolism. What is the result of performing the operations implied in $43 \div 7$, $43/7$, $1/7$ of 43, $1/7 \times 43$, $43 \times 1/7$, $7\underline{)43}$, $7\overline{)43}$? 6 remainder 1, or $6\frac{1}{7}$? When we look at real problems the answer becomes clear.

How many bags of sweets can I make with 43 sweets if there are to be 7 sweets in each bag? We use a process called *quotition*, *grouping* or *successive subtraction*. It corresponds to *successive addition* in multiplication, the process which produces the pattern $1 \times 7 = 7, 2 \times 7 = 14, 3 \times 7 = 21. \ldots$ The answer is 6, a number. It is meaningless to talk about 6 remainder 1. 1 what? $6\frac{1}{7}$ does not answer the question. Now let us look at another question: If I have 43 sweets and I wish to share them between 7 children so that each child has the same amount, how many does each child get? Here we use a process called *partition*, *sharing* or *reduction*. It corresponds to *magnification* in multiplication, the process which produces the pattern $7 \times 1 = 7$, $7 \times 2 = 14$, $7 \times 3 = 21$. . . . The answer is $6\frac{1}{7}$ though it may be a difficult task physically. Both these examples are aspects of division. How often is division only taught as sharing? How often is multiplication only taught as successive addition? They are not inverse processes. And this is only part of the story. We have not yet considered the ratio aspect or the product and separation of sets, two more ways of seeing the multiplication/ division concepts. Learning tables is useful, doing sums (if success-fully) has a tranquillising effect, but neither is a substitute for thinking mathematically or even solving simple practical problems.

I have already mentioned the possibility of seeing mathematics education as a study of relations. Let us look at something quite familiar to us. We all know and use the set of numerals 1, 2, 3. . . . What do they represent? For some they are read as a set of sounds, sometimes associated with the process of counting, some-times associated with the process of chanting. Those of us who have the latter association would have been more fortunate to have been born in that delightful primitive tribe whose counting began and ended within the set {one, two, many}. For some 1, 2, 3, . . . are symbols on which we operate in certain ways. One of the great achievements of mathematicians is that they can abstract from reality, represent this on paper as symbols, then proceed to treat the symbols as concrete objects in a new reality. But this is a highly developed way of thinking and not one used as a substitute for the process of concept development. The concepts associated with these symbols are the result of much experience with matching, ordering, sorting, measuring. In the past, we have often ignored the early experiences and restricted the study of these numbers to the useful but often sterile process of computation. Let us be selective and study the sequence known as the Fibonacci sequence, named after Leonardo Fibonacci of Pisa (c. 1170–1250): 1, 1, 2, 3, 5, 8, 13, 21,. . . We can soon find a law for finding the next term in the sequence, but the experience becomes much more exciting if we branch out into other fields of study. The initial experience may be a game of investigation, as so much of the development of mathematics has been. On the other hand it can be the result of other activity. Fertilised eggs laid by a female bee hatch into females; males are hatched from unfertilised eggs. The family tree of a male bee shows an interesting pattern. The numbers of ancestors in generations show

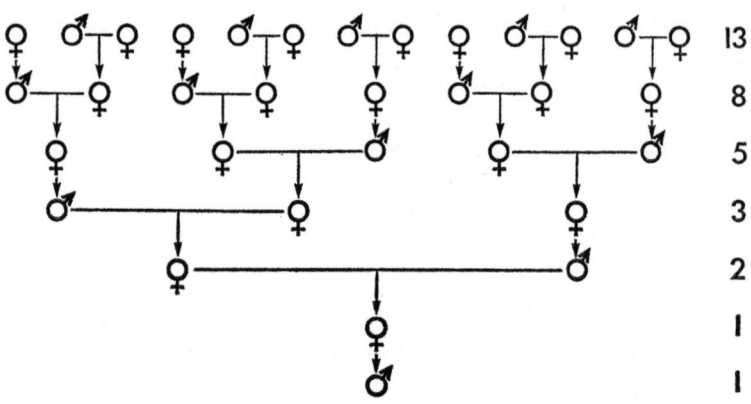

*Fig*. 2.1. The Fibonacci sequence exists in the family tree of the
     male bee.

ways in which plants grow, in the number of petals in a flower, in
the way in which leaves and shoots grow from the main stem.
    Let us look at another kind of study, involving ratios of lengths.
We can start with a rectangle with its sides in the ratio of 8/5. By
including a sequence of squares we produce a sequence of rectangles
with sides in the ratios 8/5, 5/3, 3/2, 2/1, 1/1. What have we here?
Part of the set of Fibonacci ratios. Draw a curve through the

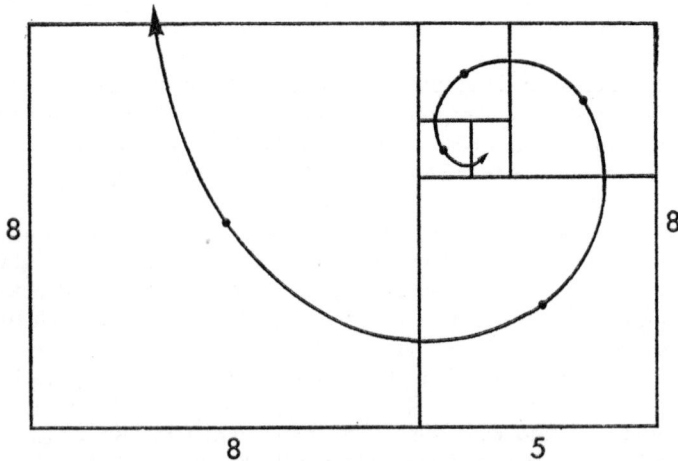

*Fig*. 2.2. Approximations to the golden rectangle.

the Fibonacci sequence. The same set of numbers appears in the centres of the squares and we have an approximation to the equiangular spiral, the spiral of the nautilus shell. What fields of study are we in now—number, quantity, space or biology? If we extend the pattern by the juxtaposition of a set of squares, 8 by 8, 13 by 13 and so on, we produce the sequence of ratios 8/5, 13/8, 21/13, . . . with a limit equal to $(\sqrt{5} + 1)/2$, the golden ratio of classical days closely associated with growth in nature and the harmonies of music, art and architecture. The architect Le Corbusier designed his buildings to be built from units based on golden rectangles because he felt, as others have felt, the aesthetic satisfaction of the proportions in the golden ratio. It was also very convenient that the sizes of his units permitted variety of arrangement. All his units were based on the golden section of the two heights of an average man, one with his hand extended upwards, one his natural height. Furniture designed to these ratios can not only be comfortable but also form a harmonious whole. What fields of study are we in now—architecture, human physiology or interior design?

*Fig.* 2.3. Furniture designers use dimensions based on golden sections of a man's height.

One approach to geometry through the study of rigid motions—translation, rotation and reflection—provides experience on which the abstractions of transformation geometry can be based at a later stage. The study of repeating patterns can be practical and orientated towards the environment. It can be integrated with design education to develop the aesthetic appreciation of geometry.

*Fig.* 2.4. A strip pattern can be constructed in seven ways.

The study of number need not be repetitive computation. A young person's interest in codes can be harnessed to the study of binary notation with the possibility of model building (Fig. 2.5).

Here we have examples of mathematical investigation which range through many aspects of human existence—scientific discovery, aesthetic appreciation, structural engineering or just plain fun. It all contributes to what makes man a thinking being, conscious of his environment. It all has its place in the education of our children.

It is the kind of study that we have been looking at that leads naturally to techniques. We start with experience (assuming the motivation for the moment) then we play, think, generalise, test the generalisation, in that order. The kind of logic used depends upon our age and stage of development but we are now sure that conventional adult logic has little value in convincing a young child of the truth of any proposition. Intuition has a much more important role. Sometimes it leads to his truth, sometimes to ours. As the investigations of Piaget showed, the number bond $2 + 3 = 5$ when describing the cardinality of two disjoint sets and their union is not a universal truth. For many children in primary school the truth of this statement still depends upon the physical arrangement of the element in the sets. If the cardinality, the number of elements,

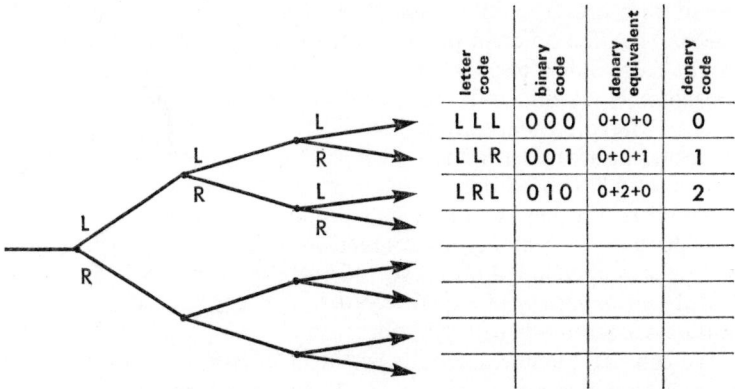

| | letter code | binary code | denary equivalent | denary code |
|---|---|---|---|---|
| L L L | 0 0 0 | 0+0+0 | 0 |
| L L R | 0 0 1 | 0+0+1 | 1 |
| L R L | 0 1 0 | 0+2+0 | 2 |
| | | | |
| | | | |
| | | | |
| | | | |

*Fig.* 2.5. A rail system can be automated for speed and efficiency.

of a set is thought to be increased by scattering over a greater physical space no amount of adult logic will convince the child of his misconception. Only the provision of more physical experience can do this, however old the child is. Similarly, if a child intuitively jumps to certain conclusions about an isosceles triangle by folding a piece of paper, it is a waste of time to try to convince him further by logic. It only convinces him that his teacher has less common sense than he has. It is this kind of approach which can destroy the excitement of the moment when a child makes a discovery—his own discovery.

It is this excitement which makes mathematical investigation so enjoyable. Sometimes the discovery is original, sometimes not, but to the child it is unique. A few years ago, a British mathematician Hyman Levy saw a link between points, curve segments and regions of a plane through studying his own doodling. In 1954 Golomb began the challenging study of polyominoes. On the other hand the tangram goes back about 4000 years and still fascinates children from 7 to 70. All these activities are now part of the normal experience of the primary school pupil. Whether the situation in the classroom is one of playing with apparently useless games or whether it is one of studying a topic of interest our aim is the same. We try to develop an awareness of structure, to appreciate it in diverse situations, to extract the mathematics and finally to develop techniques which can make further investigation fruitful and purposeful. This process is highly complex and not necessarily in distinct stages. Any investigation is subject to the influences which are already acting on the child. No two children, placed in the same situation, however carefully contrived, react in the same way. Each one starts with a different previous experience. The perception which results from immediate contact with the environment is determined considerably by all kinds of influences outside the control of the teacher. Heredity, maturation, emotional state, previously acquired concepts, previously acquired ways of thinking, ideas, expectancy, powers and forms of imagery—these all influence the child in his reactions to any new situation. Though there is an over-all pattern of development in thinking, the teacher must accept that no situation will be a mathematical one divorced from the total biological one. It will be the variety of experiences which will permit development in the greatest number of children.

We can ask ourselves certain pertinent questions concerning the kind of mathematics we introduce into our classrooms:

1. *Does it introduce the child to a rich environment?*
   Does it open up new worlds for discovery or does it repeat previous activities in the same boring way that inhibited any further interest?

2. *Does it develop awareness of relation?*
   In other words, does it develop mathematical awareness?

3. *Does it develop an appreciation of the value of mathematics in society?*
   Does the child's investigation constantly start within the environment or draw him into the environment?

4. *Does it develop an appreciation of the mathematical structure of nature?*
   Does the child's investigation involve experience with natural form?

5. *Does it develop an aesthetic appreciation for mathematics?*
   Does it produce excitement? Does it produce emotional satisfaction? Does the child have a feeling of wonder?

6. *Does it accord with modern ideas on learning?*
Does it permit all ways of learning, using all senses—aural, visual, tactile, kinaesthetic? Does it permit variety of interest and the possibility that some discoveries may not be, in the conventional sense, mathematics?

7. *Does it motivate to further study?*
Does the child often find himself thinking 'I wonder whether . . .'?

8. *Is it enjoyable?*

We can ask these questions continually. Our objective answers could guide us in our responsibilities for mathematics education. Whatever the approach to lesson organisation, whether it is time-tabled or whether it is part of an integrated day, whether it is project based or whether it is arranged through individualised pro-grammes, the experiences should take the child through all aspects of mathematics:

1. number, notation;
2. computation;
3. structure—algebraic concepts;
4. measure;
5. space;
6. representation—pictorial, graphical, geometric, symbolic;
7. language.

These are arbitrarily chosen classifications in that practice shows that many aspects can be experienced simultaneously. However, this list is far removed from the traditional diet of computation and provides opportunities for experiment and development.

In this short commentary on junior school mathematics it is impossible to discuss in detail the curriculum. As selection at the end of the junior school disappears there will be greater freedom for the teacher to provide wider experiences and encourage greater study in depth according to the interests of the children. We can here only make a plea for more experiment with courses which permit different rates of learning and different interests. If assign-ments are used they should be sometimes short to provide the satis-faction of success, sometimes open-ended to permit freedom to the child to investigate. Above all we should recognise that no child should be isolated. In our adult world, problems are solved by teams not individuals. When a child is in difficulties he must be free to do what any adult would do—consult another person; he must learn to argue, learn what it means to have a point of view. In fact the experience in our junior schools can be the foundation of future behaviour, a microcosm of adult life. In a similar way, we can see now that teachers can only produce a change in primary school mathematics if they think as a team. No one teacher is an expert in everything but every teacher can bring some special quality to a shared problem. The same is true of children. If there is any one guiding principle in all this, it is the fact that the acquisition of mathematical knowledge is not the same as learning to think mathematically.

# 3 Middle Years

## Geoffrey Matthews

Some authorities have developed middle schools, for children of 8 to 12 or 9 to 13, but the most usual pattern of education is still that children move at 11 from a primary to a secondary school. Often this transfer is a signal for an abrupt change to a formal secondary syllabus. But children don't automatically change their method of thinking from 'concrete operational' (learning with the help of materials) to 'formal operational' (more abstract problem-solving) just because they have reached a secondary school. I have met a seven-year-old who read mathematics books with evident pleasure, and who would hardly deign to touch any apparatus. At the other end of the middle years I have taught a so-called 'remedial' class of twelve-year-olds, not one of whom could count up to 30. One made a good try and got up to . . . 27, 28, but wrote the next number as 92, because that was the way it was written on the door of his new house, and he wasn't moving that one around.

The film 'Into Secondary School'* shows how two secondary schools have tried to cope with the wide range of ability in these middle years. Grass Royal School in Yeovil uses team teaching in the first two years, the children spending most of their time working on problems in small groups. The assembly hall becomes a mathematics lab for a hundred children within five minutes. Elliot's School in Putney retains classes, but the children spend some of their time on assignment cards, many of which need some simple apparatus: graph paper, scissors, string. The film illustrates two of many possible ways in which children can develop at their own rate even in the traditional primary to secondary change over. The Middle Schools represent a new attempt to help children in these middle years of 8 to 13, but there are problems: in our examination-conscious

* This is available for hire or purchase, see p. 79.

system there has even been pressure on Middle Schools to start O level courses at 11 as happens in many secondary schools now.

More important than this examination pressure itself, is the way in which middle schools are caught up in, and are between, two revolutions in mathematics education; the revolutions of *how* and *what*, of method and content. Much of the new thinking at primary level, mentioned in Chapters 1 and 2, has been concerned with how children learn. At secondary level there has been another revolution, in which a number of mathematics projects have attempted answers to the question 'What is mathematics for today?' This emphasis on *what*, on content, was quite justified when one thinks of what passed for mathematics before that time, and its effect on almost the entire population. There was a general feeling of hatred and bewilderment towards leaking cisterns, and problems such as when the hands of a clock formed an angle of $22\frac{1}{2}$ degrees.

So some secondary teachers started casting about trying to find out what mathematics would be more relevant, and from the universities, from industry and from abroad there emerged words like *sets, vectors, computers* and *statistics*, the words of the New Mathematics. These words came into school mathematics with a mystique and sometimes led to problems as bewildering and hateful as the leaking cistern problems of old. For example 'sets' seemed to be equated with sex, and work with sets seemed to consist of a search for girls with brown hair and blue eyes. This artificial exercise ended when these unfortunate girls were herded into a small pen representing the intersection of sets of 'brown-haired' and 'blue-eyed' girls.

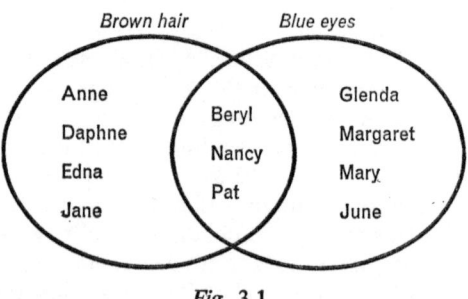

*Fig.* 3.1.

Beryl, Nancy and Pat have both brown hair and blue eyes, but so what? The idea of a *set* of objects is fundamental: indeed the whole of mathematics has been defined as 'sets with structure'. For example, there is the set of numbers {1, 2, 3, 4, ...} and adding the 'structure' of the operation of addition gives $1 + 1 = 2$, $1 + 2 = 3$, $2 + 2 =$ 4, . . . in fact a great deal of the 'mathematics' used every day. Unfortunately, in the first wave of enthusiasm, a number of people got hold of the symbolism of set theory and thought they were doing a good reform job by introducing this 'as early as possible'. In extreme cases, children were even being encouraged to write '∩' for intersection or '∃' for 'there exists' before they were even fully competent to write a 'U' or an 'E' the right way round as ordinary members of the alphabet.

There are many good reasons for 'doing sets', and one is to keep a clear mind. For example, in the past, much confusion has arisen from carelessness about which set of numbers is being used in a particular problem. Instructions of the type 'Draw a line $2\frac{1}{2}$ cm long' are misguided and misleading, as no-one has ever drawn a line $2\frac{1}{2}$ cm long and no-one ever will. The best that can be done is to draw a line as near as the eye can judge to 2·5 cm long: nearer to 2·5 than to 2·4 or 2·6. Recorded measurements, using the set of rational numbers, can only be approximations to the 'real' measure.

This may be unfamiliar thinking, even to many teachers. But much of the confusion in the past has been due to the mix-up of the different sets of numbers. For example, $8 + 9 = 17$ is indisputable and refers to members of the set of natural numbers $\{1, 2, 3, . . .\}$. But if a child draws two lines end-to-end of '8 cm' and '9 cm', the total length, even according to his ruler, will not be exactly '17 cm' and the point is that neither would it be *with the most sensitive instruments in the world*. In other words, '8 cm + 9 cm = 17 cm' is at best a clumsy way of writing a sophisticated theoretical statement and at worst a source of confusion to children.

In the bonanza of metrication, the production of measures marked '$\frac{1}{4}$ litre', '$\frac{1}{8}$ litre' and rulers marked off in '$\frac{1}{4}$ metres' are helping to perpetuate the legend of exact measurement. Again, in the past, there have been woolly examples like 'Solve $x^2 = 2$' which has no solution if working with whole numbers or if working with fractions, but has the respectable solutions $x = \pm \sqrt{2}$ in the set of 'real' numbers. Investigating the equation '$x^2 = 2$' is not exactly new—indeed it dates back to Pythagoras—but most of the so-called New Maths is really a matter of taking some of the Old Maths and making it more intelligible and relevant. The other 'New Maths' words mentioned above are shown to earn their keep in the Nuffield Guides *Shape and Size Triangle Four* (for vectors), *Computers and Young Children* and *Probability and Statistics*. All this calls for a lot of re-thinking, and as if this weren't problem enough for the 'middle years of schooling', there is the extra factor that the children are so strikingly different at this stage—some are ready to discuss the different number system and to proceed to the abstractions of algebra, others have reached a mathematical plateau and need to put comparatively straightforward mathematics to practical use rather than plunge in more deeply.

Again, there are so many cakes which we want to have and eat at the same time. Number, Quantity and Space; Concepts and Computation; Queen and Servant of Science; Pure and Applied. . . . How can mathematics be all these things? We want the children to acquire certain skills, be able to do a range of problems, but also to be able to think for themselves, to make discoveries. Sometimes, therefore, the children will be solving someone else's problem, sometimes they will be trying to devise one for themselves. One way of trying to capture the best of both worlds is the 'Closed and Open Problem'—something definite to do with an expectation that the children will carry out a further investigation, one which they create for themselves. As an example, here is a problem from a set of 52 called 'Green Problems' (each is on a separate green-backed card) produced by the Nuffield Project.

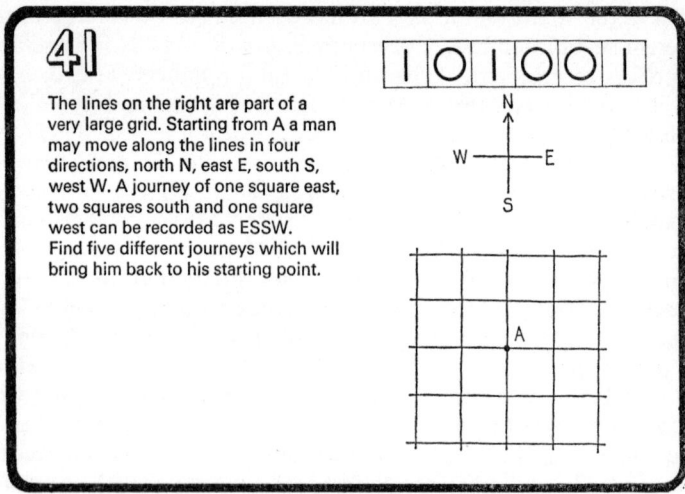

*Fig.* 3.2.

Now this problem in itself is very simple; indeed many people would vote it as downright dull. But the point is that the teacher would by this time have the children in the habit of exploring on their own. He would actively discourage a group from reporting ‹We've done that problem—there are five different journeys'. From

Plate 5
Number
square
(p. 6)

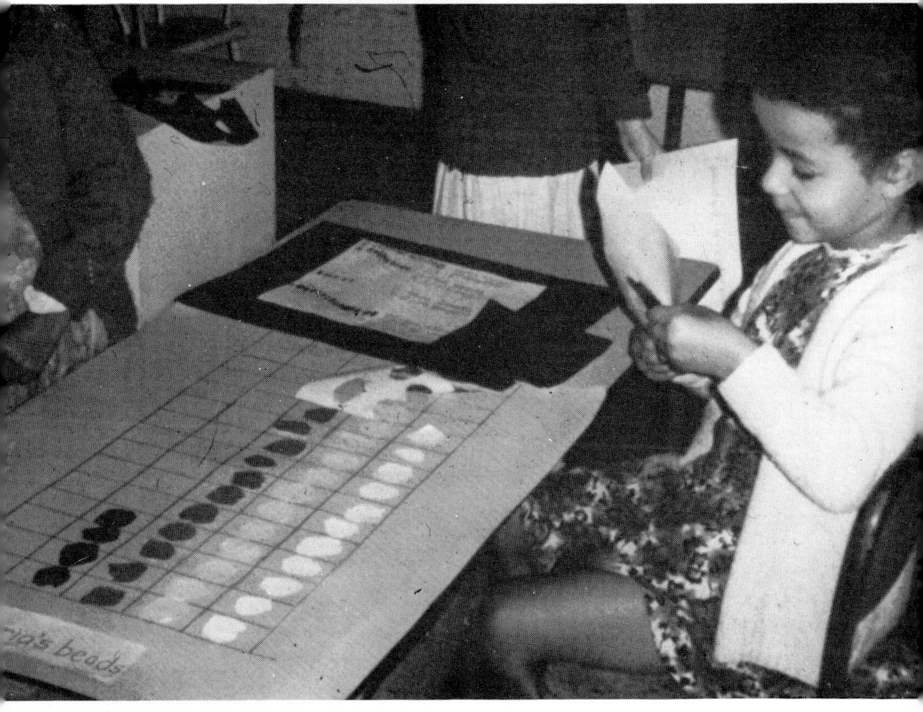

Plate 6. Recording graphically (p. 8)

*Plate* 7
Neighbour-
hood (p. 8)

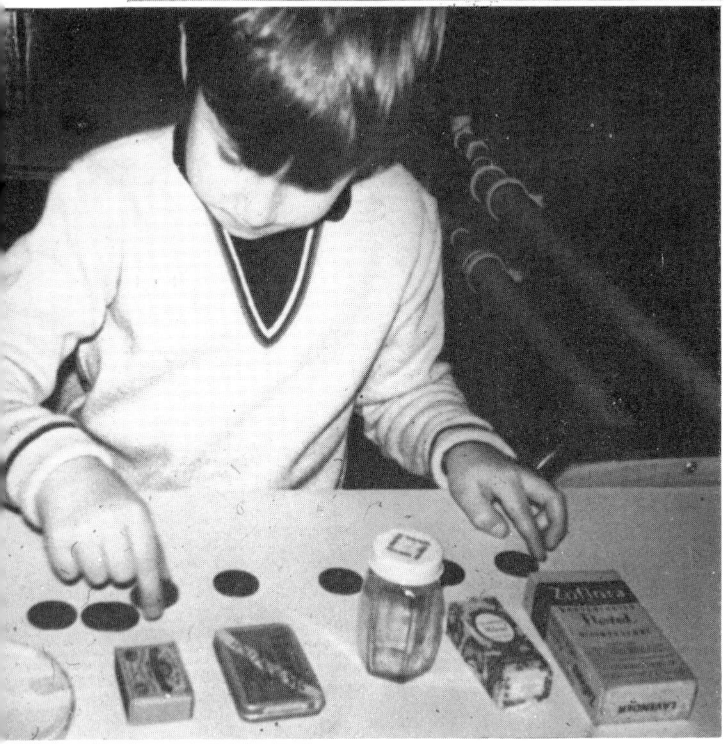

*Plate* 8
One-to-one
correspon-
dence (p. 9)

some children he might be satisfied with simply 'Please sir, we've found *six* journeys': others he would expect at least to report 'We were asked for five routes but in fact there is an infinite number of possibilities'. Others again might get tired of writing say SSSSSSS and invent their own algebraic notation, say 7(S) or S7. Once the children have got the idea of exploring, the possibilities are endless— they might draw a more complicated grid with oblique lines, for example. The previous chapter included eight criteria for the kind of mathematics to be introduced in the classroom: 'Does it introduce the child to a rich environment?' 'Does it develop awareness of relation?' and so on; I reckon this example scores 6 out of 8, which is not too bad and of course the teacher would ensure that the other two were covered presently in another problem.

The sequence 'experience, play, think, generalise, . . .' (p. 16) can at this stage be modified into a flow diagram.

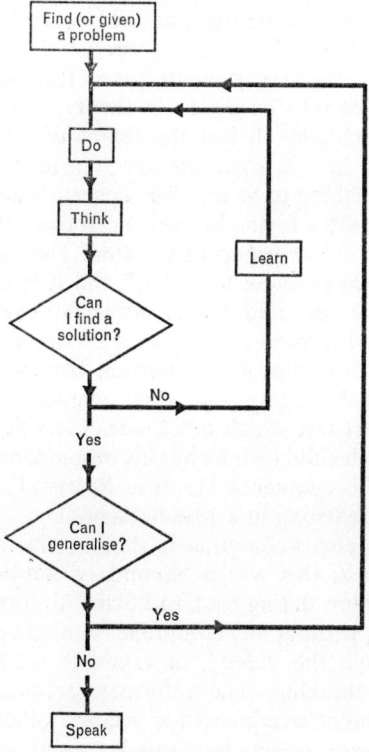

*Fig.* 3.3.

If the children *find* the problem themselves, so much the better, but in our example they were given it; they would have experimented, found the solution (5 routes) and then generalised. It is important that sometimes the children should be able to find a solution fairly quickly, for the sake of their morale, but equally important, so that they don't get self-satisfied or bored, that sometimes they should get well and truly 'stuck' and have to go round the loop to 'learn', i.e. ask a neighbour, consult a book or in the last resort get a clue from their teacher. The Green Problems have, in fact, a teachers' book which gives not only solutions but also hints of possible ways ahead, abstractions and generalisations.

A collection of such problems helps to keep the children thinking, stretched to their individual capacities, but in addition it is necessary to plan a definite progression.

Let us return to our 'concept map' of the first lecture (p. 4). There is a 'partially ordered sequence' so that, for example, the idea of similarity comes after that of congruence, but one child may be well along one branch of the tree and held up on another, while the next child may be floundering along branch 1 and forging ahead along branch 2. It is very difficult for a textbook to take such personal differences into account—in theory, it's possible to take chapter 6 before chapter 4, but the temptation to take the class straight down the line is great. In our transitional 'middle years' stage, there is something to be said for separate 'units' or modules of work, each taking say a fortnight, which can be slotted in and out to suit a particular child or group of children. The Nuffield Project is producing about 20 of these modules,* and it is hoped that other people will follow suit and devise some commercially. A more structured scheme has been devised by a group of teachers in West Kent under the leadership of Mr Bertram Banks—the children are given twelve 'tasks' ranging over their mathematical experiences, followed by a short test which determines individually the next set of twelve. Thus each child feels he has his own individual programme.

At the end of this chapter is Maureen Rowland's account of how she organises her teaching in a Middle School.

'Geometry' has also undergone re-thinking in the 'New Mathematics.' In the past, this was a 'secondary' subject and the goal was some 50 theorems dating back to Euclid. Unfortunately learning those off by heart, without any comprehension, got children through their exams, though the 'riders', or exercises, on them could lead to mathematical thinking. The reformers reckoned that 'Euclid Must Go': some went overboard for vectors, others for 'theorems without tears' (more riders, less rote-learning), others again for

* See p. 79.

'transformation' geometry. The idea is that it is not the shapes themselves which are interesting but rather what happens to them. The so-called 'isometries' are just the transformations which leave the shape and size unaltered.

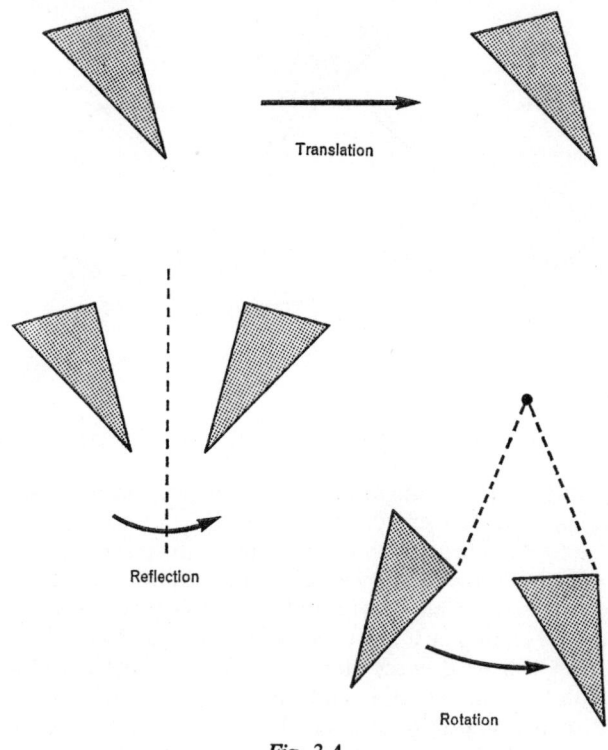

*Fig.* 3.4.

Shape and size are preserved if a shape is 'translated' (each point is moved the same distance in the same direction), reflected in a line, or rotated about a point. Are two successive translations equivalent to a single translation (*Fig.* 3.5)?

Yes. Now what about two reflections. This requires a little thought. Two rotations? A translation followed by a reflection? What about other transformations, for example enlargement? This is really a mathematical activity, asking successive questions, not leaving anything alone, questioning, generalising. If children are to be mathematical in their thinking, they must study several 'geometries', and have a battery of approaches to tackle a given problem. This again is mathematics. A good deal of 'transformation geometry' can profitably be done very early—round about the age of 8 or 9, when children are happy to discover properties of shapes literally by moving them around. Secondary texts, however, are apt to start again assuming no previous knowledge.

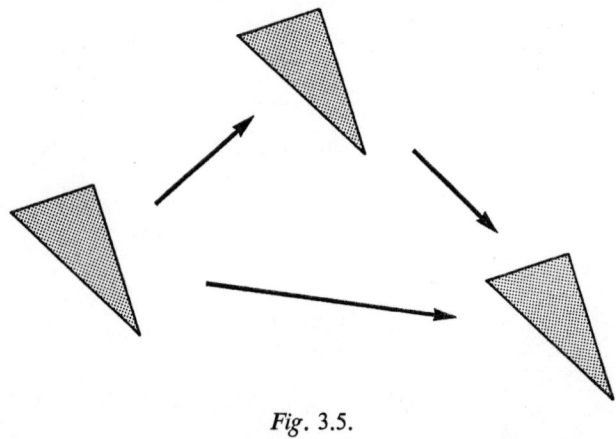

*Fig.* 3.5.

This brings up the problem of contact between schools. How, for instance, can a middle school keep in touch with the primaries and find out how far ahead the children are in their thinking? At the other end of our present scale, when the children are about 13, some will already be natural candidates for O (and even A) level. Others will be aiming at C S E (Certificate of Secondary Education), others again will need every effort and skill of the teacher if they are to go on being interested in mathematics at all. But the important message for the 'middle-years', say 9 to 13, is that children should not automatically have their mathematical way of life shattered at the age of 11 but be allowed to develop each at his own rate. Those who reach 'formal operations' early are stretched to the full and those who get there later are encouraged forward.

## Elizabeth Rowland

As each child is an individual, learning at a different rate, I try to do as little formal class work as possible. If I do need to teach a new skill to the whole class I introduce the basic concept and then leave the children to develop their own ideas and to work by themselves. We cannot expect them all to reach the same standard or to work at the same speed, so they must be encouraged to compete with themselves, and not with others. In class competition the slower learners can become disheartened, and the quicker child bored and even lazy because he has lost interest. Left alone, the children set the standard that they are capable of attaining, and they all feel successful. I feel satisfied as long as they are interested and happy while they are working, and understand what they are doing.

Mathematical games can, of course, be played by the whole class, and this kind of competition seems to stimulate interest. The children accept the fact that they need to improve their basic skills to play well, without getting the feeling that they are poor mathematicians. Knowledge of tables and co-ordinates can be tremendously improved by games of 'Fizz-Buzz' and 'Battleships'.

The main part of my teaching, or perhaps I should say the children's learning, is done through mathematical topics, which they investigate in groups. If the class can suggest a topic that we can profitably investigate I am delighted, but I usually find that I have to build up an interest first. I do this by introducing charts or models and asking the children to collect pictures or information. In this way I can stimulate interest before we start work. This may sound a little forced, but anyone who has dealt with this particular age group will know that it is not difficult to arouse interest in anything new. A rotating ring left on a window-sill is sure to be handled by one of the class, and a Moebius strip always leads to questions on topology. As soon as we have decided on a topic we discuss it and consider what mathematics is involved. This is often a suitable time to introduce a new skill that is going to be needed. I then run off assignment sheets and the children divide into groups ready to start work. They arrange chairs and tables as they need them for the lesson, and are able to clear away quickly at the end of the period. I keep separate boxes of basic equipment, and each group takes one of these and checks that everything is returned at the end of each lesson. Each box contains pencils, rulers, compasses, protractors, scissors, set-squares, measuring tape, Sellotape, felt-tips and erasers. I keep a large drawer for unfinished work, so that each class can easily find its own materials. This simplifies organisation, although the first lesson on a new topic is fairly chaotic when all the groups are collecting card and paper and looking for information.

Once the groups have settled down to work I have the opportunity to see the children individually and discuss any difficulties. By asking the right questions I try to make sure that the child knows what he is doing and why he is doing it. Some of the children can be encouraged to generalise and draw conclusions from their activities. Frequently basic weaknesses can be discovered and corrected, e.g. a child asked to find the length of the room in as many different ways as possible may not be able to use paces because he cannot multiply. If a child can be brought to see the practical use of knowing tables in this way, it is surprising how quickly he can learn. Quite often, one topic will lead naturally to another, and a child studying tessellations may become interested in transformations, symmetry or colouring problems. I always encourage this whether I have planned for it or not. Sometimes I leave the class free to work on individual projects and this has led to a great variety of subjects, including space travel, astronomy, optical illusions and mathematical games.

I think that group work is worth while, because it gives more time for individual teaching and leaves each child free to develop at his own rate. There is no doubt, however, that it needs a great deal of preparation and it is important to keep a careful check on the progress of each child to ensure that the work is purposeful.

# 4   Early Leavers

## Alaric Millington

Eighteen years ago, as a young idealistic teacher trying to build a mathematics department in one of the first two London schools to be organised on comprehensive lines, I opted to teach mathematics to the 'leavers' class. These young people, like those in similar classes up and down the country, had opted out of an examination orientated process. When we talk of ability we must be careful: performance in examinations is not the only valid criterion in assessing an individual's development in an educational process. By the examination standards of those days 'less able' was a euphemism to describe this class. Yet some had a repertoire of outstanding insight and wit, some enjoyed a remarkable amount of personal activity which kept them within the law—they had their own ideas of 'less able'. Some were better at mental computation than I was, but I had not been working part-time in the street market for years. They seemed to be a bright lot of adolescents, but they were not interested in finding the area of a tennis court, and my Pythagorean approach to the design of the confluence of a sewage system just left them suspicious of my motives. The only time I seemed to have any real influence was when the all-England champion at his weight decided to take on my six foot of pacifism, but then he was only five feet at the time.

Those children were leaving school early for many reasons; failure in the educational rat race (the feeling was more important than the fact), family pressures, personal histories and environmental influences. In our educational utopia, even if we have done all we can to keep the excitement of learning alive, we shall find in general two reasons for early leaving. There will be those who leave to work in fields of their own choice, knowing they can continue to learn. Then there are those who leave against their own wishes, in deference to the greater influence of family and peers. My leavers' class taught me that whatever we can say about organisation within the school, the other influences will often beat the best efforts of the school. I was also taught that it was educationally unproductive to assume lack of intelligence, whatever that is. In dealing with our problems we must look at the situation as it is and experiment within this, but also recognise that there is no one answer to our problems.

Three decisions made in my comprehensive school were significant in changing attitudes about early leavers. Firstly we recognised that the system of shunting them from one subject specialist to another was not in the interests of either the teachers or the early leavers. Instead, two experienced teachers whose idealism was tempered by some down-to-earth realism took over as class teachers. Regular educational visits to local industries were arranged by the teachers and the well-tried practices of the primary school were adapted to suit the ages and interests of the pupils. Visits led to projects in which pupils learned to give and to receive in joint enterprises as an alternative to the soul-destroying situation of competing with others on some artificial mark scale. Traditional subjects were used when relevant to the needs of the activities, but it needs no stretch of the imagination to think of the mathematics, English, geography, history, sociology and technology involved in a study of the local railway depot.

Secondly the leavers classes were renamed 'vocational' classes and they began to achieve a dignity and pride peculiar to themselves. We had underestimated the pride of these pupils; there was a rapid increase in the number staying after the legal age limit when there was an opportunity of taking courses within their reach and carrying some status. In the process the pupils felt they were taking some positive steps towards entering the world of work. At one time there were academic (in the narrow sense of the term), engineering, technical, craft, secretarial, clerical, pre-catering and pre-nursing courses, each leading to a goal. The 14 to 15-year-old children will leave as soon as possible if the school does not cater for and develop their ambitions.

The third decision which was significant in changing attitudes was the agreement by the staff to accede to the request for homework from those classes which had not previously done homework. Nowadays there is controversy over homework itself or its form in a modern educational system, but it was the psychology of the situation which was important. Those who felt the need for homework were asking for a recognition of their studies equal to that accorded to examination classes. The rise in mutual respect was noticeable, as was the rise in expenditure on books!

So far I have outlined experiments in one school, experiments related to the psychological implications of social organisations, and have not mentioned changing the structure of mathematics education. But it can be argued that the problems of early leavers are those which follow from social organisations, which state and confirm year after year these pupils' disillusionment. The reduction of the amount of class teaching in favour of individualised learning schedules, group activities and remedial teaching, the ending of streaming and subject setting are all directly related to acceptance of the major idea that children differ in both the rate and style of learning. The eventual consequence of these developments must be that choice of courses in the upper secondary school will follow from the pupils' motivation: selection by motivation in fact. Given adequate provision to develop and to explore, choices will then be valid. In such an ideal system it would be nonsense to describe a pupil in a pre-catering course as more or less able than a budding engineer. Each could be dangerously inefficient in the other's domain.

It is however debatable whether education should have a specific vocational bias. Some teachers—and some industrialists—would argue for a general education throughout a child's school life. However it is motivation to learn that we wish to see and at any time a child's vocational interest, or even an interest in leaving school, can be used as an aid. The Certificate of Secondary Education is not the answer. It might be if it were a certificate of experience and not of failure, or implied failure in those parts of the examination not mentioned on the Certificate. Guinness Awards and similar schemes could help in a traditional way especially if given for co-operative efforts. What can we do now about the mathematics we offer to our non-examination classes? One possibility is to accept that the application of mathematics is important. A dialogue that has been opened up between teachers and industry by the Mathematics in Education and Industry Project is proving beneficial. One student at Chelsea College has collected information on what mathematics was actually used by early leavers in industry and commerce. Research of this kind may change our ideas on what is useful. However, a strictly vocational approach is limiting if it turns education into training. If mathematics education is devoted to mastery of computational techniques and memorising of formulae, there is little that can be called creative, except when organised by an outstanding teacher. In fact the process can be boring and frustrating if it does not develop the power to challenge, permit intuitive jumps and encourage experiments with ideas. Techniques should be servants and not masters of thought. It is possible to learn techniques in solving a computational problem about a leaking bath when the sensible solution to the problem is a plumber. To consider a pupil able simply because he has learnt techniques seems strange as we generally consider a person able if he is imaginative and creative in his thinking. The changing pattern of our technological society is demanding greater flexibility in our thinking; it is transferring adults from one place to another and from one job to another. This means that we cannot know what job a particular child will take, neither can we know the future leisure patterns of our pupils. We know the prerequisite of survival in our society is the capacity to adjust rapidly in a new situation. To adjust is to look at a problem, identify and classify the elements of the situation, decide on priorities and possible routes towards a solution. This is precisely what we should try to do in mathematics. Giving pupils the opportunity to do this is part of our role as teachers of mathematics.

We can consider our role in another way. It seems to me that many mathematics courses were designed to produce a population of mathematicians. However, only a very small percentage of all our pupils become specialists. It follows that the majority of illustrations and problems we use should relate to the environment. For example, our modern approach of studying relations between and operations on sets could lead to an understanding of social structure which is more than a strictly mathematical one. Certainly we can use this study to better advantage than solving fodder problems for the British cavalry. The vast majority, the non-mathematicians, will gain a greater awareness of space if geometry is approached by a study of motion and not by exercises on bisection of angles by strictly Euclidean methods. A development of an aesthetic appreciation of pattern would encourage creative thought rather more than rote learning of formulae. Integrated studies could lead to greater transfer of training and the perception of structure and activity in the elements of the environment better than the random juxtaposition of school subjects.

It is difficult not to become dogmatic about suitable courses of action. I have outlined some experimental solutions, but they were particular to a single school with particular problems. In the last analysis teachers have to try possible solutions to the problems posed in their own school. In fact, if the suggestions for organising the primary stages were followed, firm foundations would be laid for development along similar lines at this particular secondary stage.

## Sheila Morrison

A mathematics syllabus cannot be laid down categorically; there must be latitude for what is relevant for any particular group. On a practical plane, one has to contend with staffing and discipline problems both of which can curtail some activities.

One of the topics which I have found has interested girls is a wedding: the financial side and details of etiquette, making arrangements for the ceremony and the reception. The financial, practical and legal aspects of running a car can motivate learning, as can everyday arithmetic. It is sometimes frustrating to face a demand for 'proper maths' after an energetic foray into practical mathematics. The calculating machine can arouse a long latent interest in the number system and its operation.

One should also be comforted by the realisation that a teacher can learn from her pupils; a group of boys taught me most of what I know about gambling.

## Jan Evans

At present I have only one class of girls aged 14 to 15. Some of them are leaving at Easter, others at the end of the summer term, while some are staying at least until they are 16. Those intending to stay at school after the school leaving age are generally interested in what they are learning. The others have long ago lost interest in mathematics, in fact during their first two years at secondary school. In view of this I concentrate on keeping and developing the interest of girls in their first two years. For them I have been trying to integrate mathematics with other subjects, which they like and think are easier: art, domestic science, history. For example, I am using simple recipes actually used by girls in their cookery work. This is no help to those in the fourth year whose interest has already been lost, but much of the work I try with first and second years could be extended to third- and fourth-year classes. In co-operation with other teachers complete menus could be planned: for different types of workers, for invalids. . . . Calories needed and costs could be estimated and flow diagrams of methods planned. The costs of buying and furnishing a house, credit facilities, are topics which are interesting to third and fourth years and contain a good deal of mathematics.

# 5 CSE and O Level

Geoffrey Matthews  **Introduction**

When we think of children of about 14 to 16 years old, it is rather sad that we automatically think of them as examination fodder. Poor innocent children, for a short and horrible spell in their lives they are re-named 'candidates'. 'Candidates should write on only one side of the paper', 'Candidates must attempt not more than 7 questions in $2\frac{1}{2}$ hours'. And (from an arithmetic paper in Ceylon) 'Where correct answers are given but with no working where it is necessary or with working that does not lead to the answers given, the candidate concerned may be considered as having resorted to dishonesty and punished'.

As this country is so exceptionally examination-minded, it is right to devote thought to the examinations, if only to point to some of their shortcomings and possible ways of abolishing them, or at the very least making them friendlier. Assessment is certainly essential, but must it necessarily be by the 'big bang' of the examination rather than through 'continuous creation' of confidence and ability measured by some other device, perhaps spread over a considerable period?

Before we get submerged in the examinations it is worth looking at the history of curriculum reform. Some 300 years ago mathematics began to become a respectable running-mate to the classics for a gentleman's education, and Euclid's system of geometry was accepted as a first-class training in logical thought. But the sterility of learning off by heart a large number of somebody else's theorems led, just 100 years ago, to the formation of what is now the Mathematical Association but which originated as a society for the abolition of Euclid. But things moved slowly, and in 1931 when I took my 'School Certificate' (an old man's O level), I had to commit to memory the proofs of 52 theorems. It is a sad reflection on the system that I was awarded 100 per cent on the geometry paper without having a clue what it was all about. In 1944, the Jeffrey Report suggested that Mathematics was one subject, rather than Arithmetic, Algebra, and Geometry, and that a dozen theorems were plenty to be able to prove. But it was not until 1960 that the first projects came forward with quite radical reforms for O level and fresh ideas for the newly born C S E (the Certificate of Secondary Education, designed for a rather lower range of ability). In the last couple of years, the O level examination boards have devised their own alternative syllabuses to cover the so-called 'modern mathematics', so that the rate of reform has increased dramatically.

## Tony Malpas   Secondary Mathematics Projects

If the late fifties saw the sowing of the seeds of curriculum change, the sixties was the decade when the trees grew. Of the twelve curriculum development projects in the United Kingdom sufficiently important to be listed by the Mathematical Association,* seven have been directly concerned with renewal of the mathematics curriculum as a whole for those 11 to 16-year-old children who will be CSE or O level candidates.

Table 1 gives some information about these seven projects only. Others, such as the Mathematics in Education and Industry Schools Project and the Royal Liberty School Computer Project, have been left out because they are not directly concerned with CSE or O level work as a whole, but only indirectly, or with part of an O level course.

Two points are worth making here. The first is that it seems that all of these projects were in the first instance most concerned with O level work and later broadened to include CSE. No project appears to have been started specifically in order to develop CSE materials as such. The second point is that the scale on which the seven projects have been operating and the way in which they have gone about their work varies enormously. Some projects are treating their work mainly as experiments, that is, research projects in curriculum development—for example the Psychology and Mathematics Project and the Shropshire Mathematics Experiment. Others are aiming at really large-scale reform. Two projects may perhaps be singled out in this latter respect, the Scottish Mathematics Group and the School Mathematics Project. Table 2 shows some information about these projects in more detail.

* *Mathematics Projects in British Secondary Schools.* A pamphlet prepared for the Mathematical Association and published by G. Bell & Sons Ltd. (1968).

TABLE 1. SEVEN MATHEMATICS PROJECTS CONCERNED
WITH CSE AND O LEVEL WORK.

| Project | Date started | Age and ability range | Area(s) of UK in which operating |
|---------|------|-------|------|
| Contemporary School Mathematics (The St Dunstan's Syllabus) (CSM) | 1960 | O (and A) | Various |
| Manchester Mathematics Group (MMG) | 1962 | O | Greater Manchester area |
| Midlands Mathematics Experiment (MME) | 1963 | CSE, O, (A) (11 to 18, all abilities) | Midlands |
| Psychology and Mathematics Project (PMP) | 1962 | O | Various |
| School Mathematics Project (SMP) | 1961 | CSE, O, (A) | Widespread |
| Scottish Mathematics Group (SMG) | 1963 | O and H grades | Scotland |
| Shropshire Mathematics Experiment (SME) | 1964 | CSE and O | Shropshire |

TABLE 2. MORE DETAILED INFORMATION ABOUT THE
SCHOOL MATHEMATICS PROJECT (SMP) AND THE
SCOTTISH MATHEMATICS GROUP PROJECT (SMG)

| Project | Age and ability range | Areas of UK in which operating |
|---------|-------|------|
| SMP | CSE, O, Additional O, and A. | For the Summer examinations in 1971, there were 26 000 candidates at O level. About 800 teachers attended in-service courses in 1971. |
| SMG | O grade ($11\frac{1}{2}$ to 16), top 35 to 40% of school population. H grade (16 to 17), top 10 to 15% of school population. | 600 schools 98% of O grade pupils. |

In Scotland the dissemination of the project materials to the relevant age-groups has reached near-saturation point. SMG has been a government-supported single major project for Scotland, with all the advantages, and possibly some of the disadvantages, that go with a single-minded approach to the tackling of a problem. In the more chaotic conditions to be found in England, SMP is the dominant project in terms of the sheer scale and impact of its operations. The five-year O level course books (Books 1 to 5) are all now issued and of the eight books for the Main School (CSE) course the first seven (Books A to G) are published. (Books for Additional O level and for A level courses are also published but that is not relevant to this chapter.) Each course book is accompanied by a teachers' guide. The production of twenty different books is by any standards a considerable achievement, and the standards here are high.

OBJECTIVES OF THE MATHEMATICS PROJECTS

Almost all the spade work in all the projects has been done by teachers or former teachers. In the course of doing this they have attempted an explicit statement of their main objectives. Emphases vary, but most have mentioned several or all of the objectives listed in Table 3. It is worth taking a look for a moment at each of these main objectives in turn.

TABLE 3.  MAJOR OBJECTIVES OF MOST PROJECTS

| Major objectives | Major source of pressure for change |
| --- | --- |
| 1. Refreshing and rewarding | Teachers themselves |
| 2. Relevant and realistic | Industry/business, modern society |
| 3. Renewed and respectable | Universities |

1. *School mathematics should be refreshing and rewarding*

Psychologists, indeed all of us, have long recognised the great importance of play in the lives of young children as a means by which they begin to understand their world, but it is taking us a very long time to accept the importance of its equivalent in the secondary school—learning by doing, by practical activities. Children acquire an understanding of the abstract by manipulation of the concrete. We are all now realising more and more that it is not a bit of good to expect children to sit for forty minutes working their way through twenty examples of Exercise Xh, each example differing in some subtle but irrelevant way from the one before, when interest was lost way back at Example 4 of Exercise Xa. In school maths, as in school science, material must be provided which is sufficiently interesting, novel, worthy of attention, and relevant to the real world to command the attention of children. To quote from the St Dunstan's Project: 'Meanwhile we continue to search and experiment, always looking *for more enjoyable ways of presentation* and for new topics which will be of value to the futures of our children.' This brings us to the second broad objective.

## 2. *School mathematics should be relevant and realistic*

There is also general agreement that school mathematics must start from the real world of today in which the child finds himself and that it must provide preparation for the real world of five or ten years hence when children will be starting a job.

The Nuffield Secondary Science Project (not a maths project) rates what it calls *significance* as its most important objective, and so do a number of mathematics projects: significance to the children of what they are doing in school, and significance for what they will be doing when they enter life in the highly developed industrialised society of the mid-seventies.

## 3. *School mathematics should be renewed and respectable*

There has also been general agreement on this third broad objective. To quote the School Mathematics Project: 'They [the project team] all felt that the traditional syllabuses were already long out of date and that it was *necessary for the future of mathematics in the large* that school syllabuses should be reformed.' Teachers with a concern for their subject have always wished 'to give children a basic awareness of the structure of mathematics', to quote the St Dunstan's syllabus again.

All projects are agreed on the importance of these objectives. Individual projects of course have their own emphases, but in the main all are facing the same way. The sixties might be described more as a period of curriculum upheaval, a 'big bang' to use in a different sense a metaphor used earlier, than as one of curriculum renewal. After a major upheaval it is right that there should be a period of settling down, stocktaking, getting one's breath back, beginning to see the wood for the trees. But the time is coming when we should be asking, now that we have had the big bang, where is the continuous recreation of the curriculum which will take us on a less bumpy ride into the twenty-first century?

## Margaret Brown   **O Level Examinations**

One could argue that the simplest, cheapest and most effective method of curriculum reform in this country would be to change the O and A level papers overnight. By altering their form as well as their content, one could even trigger off a revolution in not only what is taught, but also how it is taught. The publishers would rush to be first to market appropriate materials and not even the most reactionary teacher would be able to hold out against the pressure for change.

Perhaps it is fortunate therefore that the chief examiners do not wield their powers to such cataclysmic ends. In fact they rather tie their own hands by not normally withdrawing any previous syllabus when a new one comes out. Perhaps at a time when there is so much apparent activity concerned with the teaching of modern mathematics it is salutary to remind ourselves that the JMB, one of the largest and most progressive of the examining boards, still had an entry of 8000 candidates in June 1970 for their O level Syllabus A. This is the examination with three separate papers in arithmetic, algebra and geometry. The latter entails learning 25 or so theorems, which, as Professor Matthews said earlier, was condemned by the Jeffrey Report over 25 years ago. Incidentally, even though part of this figure of 8000 may be accounted for by overseas candidates, it still represents more than the total entry figure in that year for all the modern papers offered by the JMB.

Also, at least in the case of mathematics, the pressure for new O level examinations came not from the Boards but entirely from the schools in the first instance, and was the result, rather than cause, of the desire to update the content of the syllabus. After a certain amount of wrangling with the Oxford and Cambridge Board, the SMP and St Dunstan's O level examinations were approved, and both had their first candidates in 1964.

The Boards, however, were quick off the mark and by 1967 the JMB and the new AEB each had their own Syllabus C duly launched in order to challenge the monopoly of the Oxford and Cambridge Board. The other Boards all now have their own Syllabus C, or its equivalent. However, it is possible to arrange through any Board to take the special SMP papers, and a significant proportion of schools still do this. The rate of change is so rapid that it is difficult at any one time to say exactly how many schools are involved in a modern course, but we do know that in June 1971 some 26 000 candidates took the SMP O level, which was 30 per cent up on the 1970 figure.

It appears likely that at least the same number took the Syllabus C examinations of other Boards, yet the modernisation of O level still has far to go: of the 1971 entry only 20 per cent were examined on a modern syllabus.

What about the examinations themselves? In some cases it is simply new wine in old bottles. The papers are difficult to distinguish from their predecessors, except that it is now assumed that the simultaneous equations will be solved by matrix methods rather than substitution.

Some of the more enlightened chief examiners, however, have made a definite attempt to relate the papers to the spirit of modern mathematics, and have set questions designed to test the understanding of concepts and the ability to apply them in a new situation, rather than simply to test the memorising of techniques.

For example:

(1) The number of children absent from school on each day through the month of October remained fairly steady, except that on the last two days of the month there was an outbreak of 'flu, and the number of absentees rose sharply. Which of

   (a) the median       (b) the mode       (c) the mean

of the number of daily absentees is likely to be most affected by the 'flu epidemic?

$$(a) \qquad (b) \qquad (c)$$

(AEB, 1969)

(2) Three sets A, B and C satisfy the following conditions.

   $A \subset B, \qquad A \cap B \cap C \neq \phi, \qquad B' \cap C \neq \phi.$

Are the following statements true or false?

(a) $C$ cannot be a subset of $B$.
(b) $x \in B$ and $x \in C$ implies that $x$ is not a member of $A$.
(c) $A \cap C$ must be a subset of $B \cap C$.
(d) $A' \cap B' \cap C'$ must be the null set.

Answers   (a) true   false
          (b) true   false
          (c) true   false
          (d) true   false

(AEB, 1969)

(3) Place one of the signs $\Rightarrow$ or $\Leftarrow$ or $\Leftrightarrow$ if appropriate between the following statements. If none is appropriate, say so.

(i) $\triangle ABC$ is similar to $\triangle$     $\triangle ABC$ is congruent to $\triangle$
    $PQR$                                          $QPR$

(ii) $A \subset B$                                $B' \subset A'$

(*iii*) The lines $p$ and $p'$ are images of each other in the line $m$

The line $m$ bisects one of the angles between the lines $p$ and $p'$

(*iv*) $x > 3$

$(x - 3)(x - 2) > 0$

(SMP 1965)

Each of these three questions is also an example of a new type of question. Coded answers are used in the first two and short answers in the third, although this too could easily be converted to coded answer form.

The major advantages of this type of question are not only objectivity and speed of marking, but also that more questions can be set, thus testing a wider range of topics than has previously been possible.

What other innovations are we likely to see in the O level examinations, leaving aside the wider question of whether there should be two separate examinations at 16+?

There is already one instance in the JMB English language experiment of an O level grade being awarded solely on the basis of work done during the year and marked initially by the teacher. There are also a number of examinations, in particular the Nuffield A level science courses, where part of the marks are devoted to a project completed by each candidate. The ATM (Association of Teachers of Mathematics) have already formed a committee to draft an O level syllabus in which part of the assessment would be based on a series of investigations undertaken as part of the ordinary Mathematics course. Both continuous assessment and project work are already common in CSE examinations.

In the not too far distant future we may be able to tailor the assessment to suit the course chosen by the teacher rather than the course to fit the examinations.


## Margaret Hayman   CSE Examinations

Formerly the only public examination taken by school children was the General Certificate of Education, and Ordinary level was taken by about the top 20 per cent of the ability range. Partly as a result of educational re-organisation, and partly to encourage better education covering a wider range of pupils, the Certificate of Secondary Education (CSE) was introduced in 1965 for, originally, the next 20 per cent of the academic ability range. However, owing to various factors, among which are the value of an examination as an aim in a secondary school course and the increasing demand by employers for paper qualifications, the examination is now taken by pupils with a much wider range of ability. In some schools the whole ability range, except the top 10 per cent, attempt the CSE. This has made it extremely hard to devise a syllabus and set a paper which is suitable for all the candidates. This difficulty is aggravated by the decision that a Grade I in CSE is equivalent to an O level pass and is accepted as such by the majority of professional bodies. This has further increased the ability range of children taking the examination and has raised the standard of an O level pass by removing the less able groups who used to fail at O level.

The CSE examination is organised by fourteen regional Boards, and there is considerable diversity among these in their organisation and syllabuses. The examination started when 'modern maths' was new to school syllabuses and many teachers were unfamiliar with the subject. This meant that a considerable re-training of teachers was necessary. Until this could be accomplished many Boards chose to have a rather limited and more or less traditional core syllabus and several options from which schools could choose two or three to cover the diversity of knowledge among the teachers. The price paid for this great improvement in flexibility was the difficulty of maintaining comparability of standards between boards.

Of the fourteen Boards only two had a compulsory syllabus with no options. In some regions only one or two options were examined, which often allowed an unbalanced syllabus. For instance, in 1966 in one area 78 per cent of the candidates did only commercial arithmetic in addition to a very limited core syllabus. Over the country as a whole, further mathematics equivalent to an O level syllabus was the most popular option. The possible options were:

1. Further mathematics
2. Commercial arithmetic
3. Mechanics
4. Navigation and trigonometry
5. Statistics
6. Surveying and trigonometry
7. Modern mathematics
8. History of mathematics
9. Calculus and co-ordinating geometry
10. Agricultural mathematics (in part of East Anglia only).

But not all options were available in all Boards. The mathematical standards of the options varied considerably from Board to Board, some being very low and one being at least as hard as a traditional O level examination.

In addition to written papers taken at a particular time, many Boards are prepared to include, in the assessment, 'course' work done either in school or at home throughout the course, or a 'project' consisting of one substantial piece of work. The marks allowed for this part of the examination have varied between the Boards from about 10 per cent to as much as 50 per cent and, in addition, some Boards have accepted teachers' assessment of the pupils carrying up to 25 per cent of the total marks.

I think having a CSE examination has advantages and disadvantages. For the pupils for whom it was originally intended, or indeed for the 20 per cent to 60 per cent of the ability range, the examination has given an aim to the secondary school mathematics course, developing standards of teaching and attainment, and has encouraged teachers to learn something new. But where the course is followed by all the pupils except the top 10 per cent in a comprehensive school, or by all the pupils in a secondary modern school, the pace is usually too fast for the least able pupils and some of the work, while mathematically sound, cannot easily be presented in a form palatable for these pupils. The pressure of either school organisation or employers' demands forces these pupils to follow a CSE course when a more suitable one could be devised for them.

For teachers dissatisfied with the official syllabuses ('Mode 1') there is the option 'Mode 3'. This allows for syllabuses devised and examined by either a particular school or a group of schools and externally moderated. This means that an external examiner looks at the papers which have already been marked in school and decides on the marks equivalent to the various grades given in the examination.

Although the advantages of Mode 3 are obvious there are disadvantages. One of these is the difficulty of maintaining equivalent standards for several Mode 3 examinations and a Mode 1 in a particular area. Some employers have expressed doubt about a Mode 3 examination and prefer to rely on a Mode 1. Other disadvantages from the teachers' point of view are the extra time and effort needed to devise a syllabus and set specimen papers for approval by the Board up to two years before the pupils will do the examination; the cost of duplicating the papers (which is unsatisfactory) or printing (which is expensive); and the time and money spent on attending meetings with the external examiners or moderators.

Some schools have found difficulty in getting a Mode 3 recognised by a Board who are reluctant to accept it because either the syllabus or the form of the examination makes assessment difficult. For instance, one school wanted an examination with no time limit for the papers and another wanted the entire examination to consist of a

4A

project in which the part played by the individual pupil was difficult to determine. This sort of situation makes correlation of standards with other CSE examinations in the area very difficult, if not impossible.

To meet some of the difficulty of varying standards some areas, for example Stockport, have a compromise system where the pupils take one of the papers set by the Board for Mode 1 (on the core syllabus) and a second paper on an optional topic, course work or a project set and marked by the school. The performance on the first paper can be used to standardise the Mode 3 part while leaving scope for a variety of approach and interest in the school.


## Wojciech Mrozowski   Mode 3

A Mode 3 examination should satisfy the specific needs of the school running it. Thus in order to understand the contents of the syllabus and the method of examination for our Mode 3 it is necessary to see how these needs arose.

About six years ago SMP first published its O level text books. We had been dabbling with 'modern' mathematics until then, but now, using the material in these books, we were able to go ahead and construct a common core syllabus for the whole school. We quickly found that pupils of all abilities enjoyed 'modern' mathematics and were reasonably successful at it as well.

About three years ago the Metropolitan Regional Examinations Board arranged a series of meetings in various parts of London to discuss changes in their syllabus and style of examination paper. We went to our local meeting hoping to hear of changes which would provide some sort of alternative examination for those schools teaching from SMP and other similar syllabuses. The general feeling at these meetings seemed to be that the Board was changing its syllabus too rapidly. Many schools, we were told afterwards, had objected to the addition of even two or three questions on modern mathematics at the end of the traditional paper. There now seemed no possibility of an alternative syllabus or any alternative form of examination to that in Mode 1. We disliked the suggestion of a Mode 2 system which meant that we should produce our own syllabus and the Board the examination papers. We felt we should like to try some of our own ideas on methods of examination. At our district meeting there was enough interest in an alternative syllabus for us to call a meeting of those interested in exploring the possibilities of a Mode 3 scheme. The idea was to involve a number of schools in order to spread the arduous task of preparing both syllabus and examination papers. We finally started with a group of five schools, all teaching SMP and wanting some form of alternative examination papers at the CSE level. We saw Mode 3 as an opportunity to change the form of the examination as well as to write a new syllabus, but in this first phase we thought it wiser not to be too adventurous, so we submitted a syllabus based on SMP with examination papers along traditional lines.

Our new syllabus and papers were accepted by the Board without much trouble, so we then turned our attention to changing the actual form of the examination itself. We wanted to present course work with teacher assessment for 50 per cent of the marks in the examination. In order to give as much choice as possible we decided that pupils could take either two written papers or one paper plus one optional topic. Assessment in the optional topic was to be on the work produced in the pupils' folders. The examination board persuaded us, quite rightly as it turned out, that there should be not one but two options, one studied in the fourth year and one in the fifth. The options themselves were based on the work already being done in the school. We had been taking pupils out of the classroom to do some practical mathematics which involved topics such as trigonometry and simple geometry, so surveying was an obvious choice. We also had a mathematics laboratory so we thought we would extend the experimental work done there to more advanced statistics and probability. We have the ILEA planetarium on our own premises so the inclusion of astronomy was again obvious. We had recently started an advanced level computer science course and had wanted to extend computer work to younger pupils, so this was a further option. Our fifth option was the most ambitious, the making of mathematical film loops. A member of our staff had been on an audio-visual aids course at Stoke d'Abernon and had been excited by the possibilities of introducing a great deal of mathematics in the preparation and making of film loops. However, we encountered many difficulties when we attempted this and we had to give up this option after a year, but we hope to have another try at this in the future. Such were our five practical options, four of which had grown out of work already being done in the school and one being an idea which a member of staff wanted to try out.

The development of these options has had a very interesting effect in our school. Discipline problems in our fourth and fifth year mathematics classes have diminished considerably, and there are some very fine pieces of work in the boys' folders. The interest even of the form containing mainly early leavers was aroused and they asked if they too could try the options for CSE, even though they were not staying on to take the examination. The member of staff who took them decided to try the computer science course with them and he found the whole attitude of the boys improved almost overnight. We feel the reasons for this general improvement in attitude to mathematics are that they now feel that they are applying the mathematics they have learned in a way meaningful to them and that they have quite a wide choice as to which topics they follow. They feel they have been consulted about their wishes rather than forced to study something against their will.

There are two further developments we should like to make with our Mode 3. The first is a total revision of the basic syllabus. We want to add some new ideas to the initial syllabus which was based on SMP, and also to remove certain topics which we feel are not working. The second, more important part is to try and tie in our Mode 3 CSE with GCE ordinarily level. Many GCE boards have produced a 'modern' syllabus, usually termed syllabus C, which is examined by two papers. The first normally contains a number of relatively easy questions while the second paper tests deductive reasoning and mathematical thought. Our idea is that the Mode 3 examination, whether it be by two papers or one paper plus the practical options, could be regarded as the equivalent of the first paper for the GCE. So we are now trying to interest one of the GCE boards in a scheme whereby all our pupils take the basic Mode 3 examination, and those who wish to be considered for an ordinary level pass take the second paper of the syllabus C examination, the marks from which, together with the CSE marks, would determine the final O level grade awarded.

In the meantime our CSE group is flourishing. We started with five schools three years ago and now have fifteen. As yet not many of the other schools in the group are tackling the practical options. Those who are have either borrowed Wandsworth's ideas for their topics or have suggested new ones of their own. There seems to be no limit to the number of topics available—any branch of knowledge where mathematics is seen to be applied is a suitable subject. The only need is to find members of staff who are willing and able to go outside the confines of a normal school syllabus.

## Tony Malpas   **Hierarchies of Concepts**

It is one thing to formulate broad curriculum objectives, but quite another to translate these into a set of teaching objectives for a secondary school course.

It is not possible to consider this point in any detail here, but one aspect of the problem is illustrated by Table 4. This table takes four mathematical topics, two 'traditional' and two 'modern', and shows the year in which each of the seven curriculum projects, listed in

TABLE 4. YEAR OF COURSE IN WHICH SELECTED
TOPICS ARE INTRODUCED*

| Topic | Project | | | | | | |
|---|---|---|---|---|---|---|---|
| | CSM | MMG | MME | PMP | SMP | SMG | SME |
| (a) Ratio | 1 | 2 | 2 | 1 | 2 | 1 | 1 |
| Solution set of $ax^2 + bx + c = 0$ | 3 | 4 | 4 | 3 | 5 | 3 | 4 |
| (b) Finite arithmetic | 3 | 4 | 1 | 4 | 1 | 1 | 3 |
| Multiplication of matrices | 1 | 3 | 2 | 5 | 3 | 4 | 3 |

* Taken from *Mathematics Projects in British Secondary Schools* (G. Bell & Sons Ltd for the Mathematical Association).

Table 1, p. 38, introduce these four topics. The authors of the pamphlet from which this table is taken take care to point out that 'For each topic each project has provided the year in which a pupil on a five-year course would normally first meet the topic. While providing a reasonably objective criterion, this method can give no account of the depth to which the topic is taken in the first approach to it'. Nevertheless, an interesting point which emerges from the table is that projects broadly agree on the timing of traditional topics but vary more in their timing of 'modern' ones. Ratio is started in years 1 or 2 and the solution set of $ax^2 + bx + c = 0$ is begun in years 3 or 4 (or, in one case, 5), whereas finite arithmetics or the multiplication of matrices may appear at any point in years 1 to 4 (or in one case 5).

It seems, then, that where topics have been taught for a long time, the experience of teachers has produced a consensus about what

can be taught to pupils at different ages. With newer material there
is less consensus because there is less collective experience to go on.
Trials of new topics have provided some experience but not enough.
These tentative conclusions raise questions as to whether enough
work has been done on the conceptual analysis of the material
taught or on studies of children's thinking. We are, so far, not
able to state explicitly a conceptual framework within which the
various topics are contained. Work done by the Nuffield Mathematics
Project and by Piaget and his co-workers has led to the construction
of the concept hierarchy referred to by Professor Matthews (p. 2).
This basically refers to primary school children, and comparatively
little has been explicitly established, though much is intuitively
known, about the conceptual hierarchies of mathematical thinking
in secondary children. At Chelsea College we are beginning on the
analysis at secondary level, with a two-fold hope for the future.
First that secondary school teachers will be able to supplement their
intuitions and collective experience with some explicit understanding
of the thought processes of their pupils, as is happening in the primary
field. Secondly, that a basis may be established for a comparative
evaluation of secondary school mathematics projects.

# 6 A Level

## Introduction

The lectures which form this chapter were given by David Taylor, Assistant Director of the School Mathematics Project, and Stuart Parsonson, Director of the Mathematics in Education and Industry Project. Both these projects are vitally concerned with A level mathematics, indeed, as Stuart Parsonson says, MEI is almost entirely concerned with A level. There is a specific problem at this level: whether mathematics should be predominantly for the mathematically educated man in the street or for those who continue their formal education with a specialised study of mathematics at university or polytechnic. Perhaps this is not really a conflict. The demand for increased numeracy among professionals of all kinds leads to more mathematics education beyond O level; this requires more mathematics teachers which in turn involves an increase in the numbers reading mathematics in higher education. Both projects have attempted to resolve the apparent conflict. The accounts also illustrate the general problems of curriculum renewal, the additional topics to be included in the syllabus, how few old topics can be left out, the need to find new methods which prevent overloading of pupils.

Both lectures were built around questions posed by Margaret Brown, who introduced the speakers. These questions were on the content of the syllabuses, and especially the inclusion of new topics in pure mathematics, the form of examinations set, the unforeseen difficulties in working out the project, the present position of the project in the school and plans for the future. The first question posed to both was 'How did the project begin?' Both speakers referred to the 1957 Oxford Conference arranged by Dr Hammersley. This conference was perhaps the first time school mathematicians, university mathematicians and industrialists met together. At this conference and at those which followed in 1959 and 1961, it became clear that mathematics teaching in school was in trouble. Obvious sources of difficulty were the gaps between mathematics in school on the one hand, and mathematics in universities and in industry on the other. In their different and independent ways the School Mathematics Project and the Mathematics in Education and Industry Project attempted to close these gaps.

## David Taylor  School Mathematics Project

After the conference of Summer 1961, Dr Thwaites got together a group of four heads of mathematics, T. A. Jones from Winchester, D. A. Quadling from Marlborough, T. D. Morris from Charterhouse and H. M. Cundy from Sherborne. Their objective was to write a completely new O and A level syllabus with the necessary texts, starting with pupils of the 13+ age-range, as that is the age of entry into public schools. It was felt at that time that probably only the public schools had sufficient independent control over their work to undertake this kind of exercise. The acceptance by the universities of a new A level was crucial and such a group of schools was in a good position to ensure this. The team then set about the task of writing the texts and preparing the syllabuses and of persuading every university mathematics department and every Oxbridge College that they could only gain by accepting pupils who had taken this new A level course. At the same time the Oxford and Cambridge Schools Examination Board agreed to provide a special examination for the pupils who had followed this new course. Later, other boards were persuaded to allow pupils from their schools to take the same papers, an arrangement which still holds. The original objectives were twofold: one was to improve *what* was being taught in schools and *how* it was taught, and the second, dependent on it to a certain extent, was to increase the supply of mathematicians. We certainly see an improvement in the number of pupils staying on to read mathematics in the sixth form, but we have no firm statistics to know whether the number of mathematicians in universities has increased.

It follows from the original objectives that the SMP syllabus will certainly differ from traditional A level, but of course will have many things in common with it. We have included a great deal of work on both algebraic structures and vectors; indeed our approach to mechanics is completely vectorial. The introduction to the calculus is treated in a different way from the traditional approach of gradients. I am not sure how successful this has been, as we have had very little feedback on this, but we use the idea of an average scale factor and after working for some time with this approach, we show how this is equivalent to finding gradients of curves. We have also introduced electricity into the course; for SMP Mathematics at A level, only a knowledge of d.c. theory is required, but a rather more advanced treatment is given, including a.c. work. These are extended in the Further Mathematics text. Statistics and probability are also developed to a high level. I suppose, having said that we have included all of these topics, I had better mention what some people say we have omitted, namely statics. But I think that the general view of SMP is that statics is a subset of dynamics and any statical problems can be treated as such. The underlying reasons for making these changes are to make mathematics more meaningful and relevant to the experience of the pupil and to his needs when he leaves school.

We have therefore tried to show the physical applications of abstract algebraic structures wherever possible, indeed introducing these in the O level course, where we first meet groups (without naming them as such). In the sixth form, the course for a single A level gets up to fields, working successively through the various structures. In Further Mathematics, work on vector spaces is included. This is obviously preparation for the university pure mathematician. But abstract algebraic structures evolved as a result of someone trying to solve physical problems and therefore have relevance to aspects of engineering, industry and other disciplines.

Our intention was to make mathematics a single A level subject. However, from the beginning there were in fact two examinations, Mathematics and Further Mathematics, partly because of pressure from a number of the original SMP schools. The form of the examination has changed considerably since its first setting in 1966, and we hope that we now have a more realistic situation in Further Mathematics which perhaps I might outline. The first exam on the new arrangements of the Further Mathematics syllabus took place in 1971. This is now split into five blocks rather like an English Literature paper. These are (i) linear algebra and geometry, (ii) vectors and mechanics, (iii) differential equations and circuit theory, (iv) extensions of calculus, (v) statistics and probability. Most candidates only need to have studied four of those.

Perhaps I should explain how the single subject SMP Mathematics is examined at A level. For a number of reasons we changed the original pattern two years ago so that the first paper, still lasting three hours, now consists of twenty-seven short questions, of which eighteen correct solutions are required to obtain full marks. The second paper consists of ten reasonably long questions of which candidates have to answer seven to get full marks. So the two papers are completely different in layout but are equally weighted in the overall result. There is also a Special paper to go with this exam.

For Further Mathematics the first paper consists of four short questions on each of the five blocks I have mentioned, and candidates have to answer a total of fifteen questions to get full marks. The second paper consists of two questions in each of the five blocks, and the candidates have to answer five in all to obtain full marks, but again the two papers will be equally weighted. None of the questions are of the objective type but, on the whole, each of the short questions concentrates on one idea. We are not sure that we have got the right way of doing it yet and we are always prepared to consider other ideas and other viewpoints.

Both A level syllabuses have already undergone two stages of revision. The balance is perhaps the major difficulty and I don't think anyone can say exactly what it should be. It's just a matter of finding out by discussion, by seeing what people want, by sensing what is really appropriate. We now see that the first two A level books are not very satisfactory, as they don't follow on well from the O level books and are not a very good preparation for a complete A level course. Part of the problem, in schools which are recently adopting the SMP texts, is that they get hold of the book and the teacher learns Chapter 1 in Week 0 so that he can teach it in Week 1. This means that we really must try to get the best order for the chapters, for this sort of teacher. In the existing books I don't think we've got the right order; for example, calculus comes much too late, so if you follow the order of the chapters you get the physics master complaining that they can't differentiate by Christmas, and the chemist complaining that they can't integrate by spring. We are therefore re-arranging and re-writing the whole lot. This takes a long time because we believe in testing our materials before they are published.

As to the numbers of candidates and of schools involved, I am never sure which statistic to use to show the way in which SMP has succeeded. For example, we know that SMP texts are in use in well over a thousand secondary schools in the UK and this represents somewhere between one sixth and one quarter of the secondary school population. We know they are used, but we don't know in what way or at what levels, as these are merely figures that we get from the publishers. The entries for SMP exams can give some measure. For example, last summer in single subject A level there were nearly two thousand candidates and for Further Mathematics, two hundred. At O level, as a matter of comparison, there were just over twenty thousand in 1970, but many schools use our books as preparations for syllabus C. This applies especially at O level, but increasingly it will also happen at A level, where JMB now have a syllabus C and I believe London are preparing one, while others take the MEI examination. Indeed some people use other text books for our examinations, so it is altogether an unreliable statistic.

The A level text books 1, 2, 3 and 4 have now been out for some time, and they have now all been metricated. The five Further Mathematics books are now available, one for each of the blocks in the syllabus. There have been draft texts which have been privately produced and circulated to schools over the past five years. We also have supplementary handbooks, 'Statistics and Probability' written by John Durran and 'Calculus and Elementary Functions', which is an alternative approach, by Jones and Montgomery. These are both under the SMP publications. For the really broadminded teacher who wishes to pursue his modern ideas to a deeper level, we are preparing 'companions' to advanced mathematics. The first one, already published, deals with algebra and analysis. It looks at what we have done in our advanced course and projects it into the university one, attempting to tie up the two. Volume Two, dealing with Statistics, is well under way. Apart from the bread and butter rewriting and filling in of gaps, our immediate planning is concerned with computing, but we are all the time considering how we might prepare ourselves for the re-structuring of public examinations. No-one seems to know what form this is going to take; perhaps O level will vanish or be replaced by something else. But we are trying to be ready for whatever might come and are in fact thinking that we might have to re-build the project completely and start from 9+ and go to 16, 17 or 18. Who knows?

## Stuart Parsonson    **Mathematics in Education and Industry Project**

The origins of MEI stem from the 1957 conference, but the true beginning came in 1962 when B. T. Bellis, who was then the senior mathematics master at Highgate school, spent a sabbatical term at Oxford studying the relation between mathematics as taught in schools and the mathematics used in industry. When he returned to Highgate he gathered together a number of schoolmasters from schools in North West London and suggested setting up a group which would look into the specific question of liaison between schools and industry. From this beginning has developed a scheme under which schoolmasters go into industry and industrialists come and speak to schoolmasters, and this side of MEI activity has continued to the present day. There are something like a dozen MEI groups around the country supported by the Industrial Committee of the Mathematical Association which are looking at the specific problem of communication between schools and industry.

The replies, when we asked industrialists what they thought of school mathematics, were unflattering. This was as expected; if you ask the secondary school master what he thinks of the teaching in primary schools the reply tends to be unflattering; the university lecturer would be equally unflattering about teaching in secondary schools. This is an inevitable result of the lack of communication. However, there were some criticisms voiced by industrialists which we thought were based on substantial evidence and were well informed. There was the criticism that workers on clearly routine tasks in the mathematical sense of the word were unable to do simple arithmetic; they didn't know the multiplication tables for example, and when faced with a decimal, they became utterly incoherent. Then there were those who were supposed to show a certain amount of initiative in their work who lacked what was described in those days, somewhat vaguely, as numeracy—a word which has come to mean something rather more definite now. For example, if faced with a summary of numerical data in graphical form they were unable to read it intelligently, or even to interpret it. Then more surprisingly there were criticisms of people who came in at the highest grades: people with degrees in non-mathematical subjects who had no conception of the meaning of a mathematical argument or were unable to assimilate numerical data and make decisions. Most surprising of all, many entrants to industry with technological qualifications and with mathematical degrees were incapable of solving the mathematical problems which they met in industry. One man in charge of recruiting at a very large firm said that they had eight entrants with first class mathematical degrees, and of those eight only one was of any use to the firm; the other seven were totally incapable of using mathematics in an industrial context. We realised that here was a common problem: being able to abstract from physical situations a mathematical model. To be able to do so is a rare gift, but it did seem that schools were insufficiently aware of the problem. So round about 1963/64, although MEI had originally been concerned with liaison between schools and industry, it seemed necessary to devise a new mathematical syllabus, hence the present MEI syllabus. This started in a rather different way from SMP. Because of the diversity of schools which had become interested, the syllabus was aimed at the higher age-range, rather than the lower one, so that it would affect fewer members of staff to begin with. Essentially MEI is an A level scheme which has not attempted to develop an O level elementary stage. A revised additional mathematics syllabus (at O level) was produced in 1965 as an interim measure and has continued successfully ever since.

Although the project is called 'Mathematics in Education and Industry', the Education and Industry is non-commutative; education comes first at all times. No subject has been introduced into the MEI syllabus specifically because of its industrial uses; they are all there because of their educational value. Industrial aspects of mathematics were, however, kept in mind in devising the syllabus.

The major changes in the syllabus were, I suppose, a considerable reduction in the amount of technique and a major increase in the amount of mathematical model-building required, particularly in the building of differential equations and the introduction into A level of a significant proportion of probability and statistics, which was a rarity then, though it is rather more common now. We strongly felt that mechanics had a place in the syllabus and was in danger of erosion. We thought that, while keeping mechanics, we would approach it in a somewhat different manner; indeed we felt this was an admirable opportunity to bring in the concept of mathematical model building. The mechanics questions we set are standard ones on Newton's equations of motion, and momentum and energy, but they look a bit different. Candidates are required to abstract the mathematical models from the physical situations presented and then to solve the resulting problems. The questions can often be solved in two or three lines: they are mathematically very simple from the point of view of technique required, but they are quite demanding in the sense that they examine the concept of building a mathematical model. We feel that here we are giving a new look to mechanics. In a sense we have gone some way towards rejuvenating mechanics as a subject.

There was also the attempt to devise a more coherent syllabus. To take specific examples, there was the calculus thread which ran through a whole lot of the work including numerical analysis, and what might be described as the vector thread which runs through the whole pure mathematics course. A considerable amount of traditional mathematics was retained; in fact some people might say too much. For example, much calculus is still there, but the emphasis is rather different because more importance is attached to arriving at a numerical answer. One surprising criticism industrialists made was that people were quite happy to solve mathematical problems containing data in algebraic form but to give them numbers was to ensure that they got an absolutely absurd result.

In algebra the emphasis is on linear algebra, although that is a rather formidable term for the idea of using matrices and vectors effectively in various situations. We have introduced in the syllabus two or three lines indicating that we expect candidates to have some knowledge of algebraic structures, but this doesn't really require axiomatic systems; we simply expect them to know what a distributive, commutative or associative law is. We felt that knowledge of actual algebraic structures was certainly valuable for those who studied mathematics beyond the sixth form, but that it was not immediately required by most users of mathematics and the vast majority of sixth formers are in this second category. We regard algebra at this level as a tool for getting on with a job, so our emphasis is on matrix and vector algebra as the most important application for the mathematically educated man in the street. In the S level double-subject Pure Mathematics we include linear algebra in a rather more abstract way including the axiom sets for fields and vector spaces, but this is very much top level stuff.

It is interesting to observe how close MEI S level Pure and SMP Further Mathematics are, for it was on this point of double subjects that MEI differed from SMP. The original aim of SMP was a single subject examination, and the original MEI schools felt very strongly that although single subject mathematics ought to be the central mathematical examination, there was a significant and very important group of students in the schools at all times who were simply not content with doing single subject mathematics. They were too good for it and ought to be given a more challenging course; we therefore stuck out for a double subject in mathematics. Effectively SMP have almost come round to the same point of view, because they have a Further Mathematics examination. SMP were probably right to go for the Mathematics and Further Mathematics format: this makes more sense than Pure Mathematics and Applied Mathematics, where it is always desperately difficult to draw the line. For instance, a whole lot of calculus comes into the Applied Mathematics because there is too much that we want to put into the Pure Mathematics. The combination of Mathematics and Further Mathematics makes better sense, but there is one drawback: you get a bimodal distribution of examination marks. Mathematics is one of those subjects where you have some exceptionally clever people who distort the examination for the vast bulk of candidates. On the whole the SMP approach of getting everybody to take the single subject and then pushing the others through further papers has much to commend it, but it has this one limitation we felt should be avoided. We might change our approach in future, however.

When we came to devising examination papers we felt that since we had changed the content, perhaps we should leave the format essentially alone, and this has been done throughout the pure mathematics papers where, say, seven from ten or eleven questions are attempted. It is slightly different in applied mathematics because we have re-interpreted applied mathematics to include not only the traditional mechanics but also statistics and probability, for those whose interests lie outside the physical sciences and are concerned with subjects like economics. We have split the applied mathematics papers into a mechanics section and a probability and statistics section, but we also have a short compulsory section on probability at the beginning of the paper, because we felt this was so important that everybody doing mathematics ought to know something about it. At S level we begin with a short section of what might be described as harder A level questions, and then the rest of the paper is set on 'topics'. The idea here is that a student good enough to go on to S level is able to read on his own a specific topic and then answer examination questions on it. This is rather an idealistic point of view; I think most schools in fact teach for the topics, usually choosing two out of three, but a pass or a classification can be obtained on the first section and on one of the topic sections of the paper.

Perhaps with the benefit of hindsight it is possible to see more clearly the difficulties in devising a new syllabus. The first point is that in cutting out techniques there is a very real danger that you replace them by other techniques: you can derive techniques for solving questions on group theory, just as easily as you can derive them for solving questions on the geometry of circles or triangles. Indeed, you can't really get an examination syllabus without having some body of techniques there in the long run: difficulties arise if you try to cut down their number. For example, there always used to be a course on what was called systematic integration, which meant starting with the integrals of very simple functions and gradually building up the number of standard forms. We tried to cut this down but it has not been a complete success; pupils are not quite as good at integration as we would like. They have to practise varied and sometimes taxing integrals if they are to recognise standard forms and become effective at doing integration. If a technique is watered down to avoid unnecessary labour it might be watered down so far that it can no longer be applied. Another example is in solving trigonometric equations; pupils used to solve unnecessarily hard trigonometric equations, but it is quite useful to be able to find out those angles between $0°$ and $360°$ which satisfy the equation $\cos x = -\frac{1}{2}$, for example. Cutting down the amount of practice at manipulating trignometric equations may have the result that pupils are incapable of answering even such simple questions. There is a very delicate balance—cutting down the practice of techniques in the end may not result in genuine time saving. Another difficulty that we have found, and this is a major difficulty, is that statistics at school level has proved not to be comparable intellectually with mechanics. We have looked carefully into this; the major part of our statistics syllabus is mathematical probability, which we have made quite exacting. Even so, pupils still choose it rather than the mechanics. The underlying concepts of mechanics are very deep and subtle and they take an awful lot of time to master. It's another question really of watering down; if you water down mechanics too far students aren't capable of answering anything on the subject. Another topic that we tried to retain was projective geometry because we thought that this was one of the most beautiful aspects of school mathematics. This unfortunately has not been a success; I think it will inevitably be dropped as a topic in the S level after a time: pressures from outside do affect the syllabus and in these rushed and hurried days people seem to have no time for this elegant subject. We have fought to save it but have failed. Another point in geometry is that we have attempted to retain some knowledge of the conic sections, partly because of their very considerable historical interest, and partly because they are very useful: they come up in applications time and time again. Here again one has to be very careful to keep a balance; it is no good just introducing the standard forms of equations of the conic sections, finding the equations of tangents and normals and hoping that this gives any sort of coherent picture. A study of conics must include some of their geometry. The saddest thing of all is that the new mechanics that we set is interesting to teach and wonderfully stimulating for classroom discussions, but not, it seems, an ideal examination subject in its present form.

Despite these difficulties I still think that the MEI examination is a significant improvement on traditional mathematics examinations. I enjoy teaching it more and I think that we have a more lively and flexible examination at the end.

One indicator of success is the way in which schools take up our syllabus. About 40 schools took the A level examination in 1970 and this should reach 80 by 1972. Of candidates, about 1000 took single subject Mathematics, 300 took Pure and about 300 took Applied Mathematics. From the number of enquiries received there is likely to be a considerable increase in the future.

The policy of not publishing texts diverges from that of many projects. We felt that the first priority was to define the syllabus rather than run the risk that the contents of prescribed texts would dictate the content of future examination papers. For a while we did circulate to schools cyclostyled notes on topics, but now a number of books are appearing which enable teachers to choose the most appropriate text for them and their pupils. MEI does not wish to make any selection of texts which define a rigid syllabus and teaching style for those taking up MEI.

Flexibility in syllabus and examination is our major concern and in the future we shall simply see where events lead us. For example, our major experiment in the coming years is the use of work produced other than during examination time for assessment at A level. It is difficult to ensure comparability, but we obviously must encourage pupils to do statistical field surveys and to write computer programs for the problems which interest them. Already work of this kind has been submitted; some of it is of such good quality that we hope to see much more in coming years.

# 7 Breaking down Barriers

## Introduction

The previous chapters have given some indication of how mathematics in school has become less insular and less artificial. Two factors will help to accelerate this process: the advent of the computer and the increased demands for mathematical sophistication in other subjects.

In the first section of this chapter, Barry Blakeley points out how computers will become more and more available to schools and how access to them will enable children to work on meaningful problems which were inaccessible in the past because of the sheer tedium of the calculations. Bob Lewis in Section 2, on science in general, takes up this point in discussing how the numerical solutions are obtainable by computer, this again permitting the teaching of real applied mathematics to scientists.

In the following sections Margaret Brown discusses the possibilities of joint 'modules of work' linking mathematics and science and also gives an example across the boundaries of Mathematics and Sociology.

Finally Brian Dudley describes two pieces of work in Mathematics and Biology.

## Barry Blakeley  Computing

I am convinced that children must be educated about computers from an early age. The 'early age' can be as young as nine or ten; consider the work done by Desmond Brighton in the Nuffield Mathematics Project. I feel that it is essential for every pupil, by the age of fifteen, to have had a thorough grounding in the history of calculating devices; how the computer works; what the computer can do (including the writing and running of a number of programs in a suitable high-level language); what the computer is doing in industry and commerce, and hence the commercial and social implications of the computer. In the years to come, computers will be available in ever increasing numbers in schools. They will affect many school subjectsi including art and music, geography, economics, the sciences and mathematics. They are going to change the approach to these subjects because they are going to make possible techniques that until now have been impossible and they are going to provide information that until now has been unobtainable in the classroom. The seventies will be a most exciting time for teaching.

But now I must turn my attention to computing in mathematics. If pupils have been receiving the sort of education I have already mentioned, computers will have affected their mathematics long before they enter the sixth form. The presence of a computer in the classroom necessarily alters the way we approach the teaching of mathematics.

The basic requirements of programming are every bit as valuable in the sixth form as they are lower down the school. Flow-charting and the writing of programs emphasise the two parts to the solution of any problem: the definition of the algorithm (the set of precise unambiguous instructions which must be followed to obtain a solution) and the factual information of our problem (the raw data that must be combined with the algorithm to give the appropriate solution). Somebody once said that one cannot say that one understands a piece of mathematics until one can go into the street and explain it to the first person one happens to meet. I have always felt this to be a counsel of perfection. After all, the first person one meets may be an idiot, but in the computer we have the perfect idiot! Perhaps we cannot say that we understand a piece of mathematics until we can program a computer to use it. All this is of the greatest value in training in logical thought at any age. Here we have a modern equivalent of Euclid. We have a fixed set of allowable instructions and using only these we must produce a logical sequence to solve a given problem. As we progress we create structures, somewhat akin to Euclid's theorems, which are building blocks that we can use in writing more complicated programs. All this is built up on a strictly logical basis. You may care to produce a set of instructions, using only $A = B + C$, $A = B - C$, $A = B \times C$ and $A = B \div C$, to find the square root of a number to any desired accuracy. There are a number of ways of doing this.

In the lower school using the computer can help in teaching the importance of the order of operations. The idea of nested multiplication can teach economy of calculation as well as test a pupil's

manipulative ability in algebra. It is possible to investigate, on a
practical basis, sequences and series and the idea of convergence
soon begins to appear, although we are not *proving* that a particular
sequence converges. Recurrence relations can be investigated at a
simple level. An approach to the calculus from these beginnings has
much to recommend it. It is a most valuable exercise to see the
gradient of a chord approaching a limit as one end of the chord is
made to approach the other (fixed) end. Pupils can be encouraged
to perform their own investigations. Who can say what results may
be produced by such activities, and what properties discovered?

In the sixth form all previous uses can be continued and extended
as new topics are introduced; for example, iteration techniques have
received little attention in school mathematics, but they are of the
greatest importance in writing computer programs. Suppose the
problem is to find the square root of 5. The sort of approach used
by a fourth former who has left his tables at school and needs to
know the value of $\sqrt{5}$ for his homework is that $2^2 = 4$ and $3^2 = 9$
so that result is between 2 and 3. He tries 2·1, squares it and as the
result is smaller than 5 tries 2·2, squares it and so on until the
square of 2·3 is larger than 5. At this stage he goes back to 2·2 but
increments by steps of 0·01 until the square exceeds 5 and then uses
steps of 0·001 and so on until the result is as accurate as is required.
It is not a difficult matter to write a program for this and the results
of a typical computer run are given in *Fig.* 7.1.

This sort of work is, of course, more appropriate to third and
fourth formers, but in the sixth form it could be followed by more
sophisticated methods leading to the Newton-Raphson method and
the idea of quadratic convergence.

Another topic to which the computer lends itself well is that of
differential equations. It is interesting to approach these via numerical
solutions and then to deal with the analytical approach. This is the
method used in the *School Mathematics Project Advanced Mathematics, Book 3*. Given the information in the table in *Fig.* 7.2, about
the acceleration of a car at various times after it has started from
rest, we wish to find out how the velocity of the car varies with time.
There is no alternative to using a numerical method.

Replace the average acceleration over a small interval of time by
the value of the acceleration at the beginning of that interval of
time and calculate the velocity at the end of the interval from a
knowledge of the velocity at the beginning of the interval, using the
fact that

$$\delta v \simeq \frac{\mathrm{d}v}{\mathrm{d}t} \times \delta t$$

*Trial and error method*

2·1
2·2
2·3
CHANGE INCREMENT TO ·01
2·21
2·22
2·23
2·24
CHANGE INCREMENT TO ·001
2·231
2·232
2·233
2·234
2·235
2·236
2·237
CHANGE INCREMENT TO ·0001
2·2361
CHANGE INCREMENT TO ·00001

and so on.

*Fig.* 7.1.

| Time | 0 | 2 | 4 | 6 | 8 | 10 | 12 | (in sec) |
|------|------|---|---|-----|-----|-----|------|----------|
| Accn. | 11.25 | 9 | 6 | 5.6 | 4.9 | 4.1 | 3.75 | (in ft/sec$^2$) |

*Fig.* 7.2.

The time interval will be 2 seconds. Then we can use the calculated value of the velocity as the starting value for the next interval of two seconds. This way we obtain the following results:

| Interval | Approx $dv/dt$ | Approx $\delta v$ | $t$ | $v$ |
|----------|----------------|-------------------|-----|-----|
|          |                |                   | 0   | 0   |
| 0–2      | 11·25          | 22·50             |     |     |
|          |                |                   | 2   | 22·50 |
| 2–4      | 9              | 18                |     |     |
|          |                |                   | 4   | 40·50 |
| 4–6      | 6              | 12                |     |     |
|          |                |                   | 6   | 52·50 |
| 6–8      | 5 6            | 11·2              |     |     |
|          |                |                   | 8   | 63·70 |
|          |                |                   |     | etc. |

Fig. 7.3.

An improved method would be to take the value of the acceleration at $t = 2$ to represent the average value over the interval from 1 to 3 seconds, the value at $t = 4$ to represent the average value over the interval from 3 to 5 seconds and so on. We must use the first method, however, over the interval 0 to 1 seconds in order to obtain a value of the velocity at the start of the interval 1 to 3 seconds. Using this improved method we obtain the following results:

| Interval | Approx $dv/dt$ | Approx $\delta v$ | $t$ | $v$ |
|----------|----------------|-------------------|-----|-----|
|          |                |                   | 0   | 0   |
| 0–1      | 11·25          | 11·25             |     |     |
|          |                |                   | 1   | 11·25 |
| 1–3      | 9              | 18                |     |     |
|          |                |                   | 3   | 29·25 |
| 3–5      | 6              | 12                |     |     |
|          |                |                   | 5   | 41·25 |
| 5–7      | 5·6            | 11·2              |     |     |
|          |                |                   | 7   | 52·45 |
|          |                |                   |     | etc. |

Fig. 7.4.

Having introduced the methods in this way it is advantageous to compare the results obtained by such methods with a known, accurate solution.

The results of these methods for differing step lengths are shown in the table in *Fig.* 7.5, together with the known solution of the equation

$$\frac{dv}{dt} = 3t^2 + t + 1$$

given that $v = 0$ when $t = 0$.

| | | Method I | | | Method II | |
|---|---|---|---|---|---|---|
| | | step = 1 | step = 0·5 | step = 0·1 | step = 1 | step = 0·5 |
| *t* | *true v* | *v* | *v* | *v* | *v* | *v* |
| 0 | 0 | 0 | 0 | 0 | 0 | 0 |
| 0·5 | 0·75 | | 0·5 | 0·69 | | 0·719 |
| 1·0 | 2·5 | 1 | 1·625 | 2·305 | 2·25 | 2·438 |
| 1·5 | 6·0 | | 4·125 | 5·595 | | 5·906 |
| 2·0 | 12·0 | 6 | 8·75 | 11·31 | 11·5 | 11·875 |
| 2·5 | 21·25 | | 16·25 | 20·2 | | 21·094 |
| 3·0 | 34·5 | 21 | 27·375 | 33·015 | 33·75 | 34·313 |
| 3·5 | 52·5 | | 42·875 | 50·505 | | 52·281 |
| 4·0 | 76·0 | 52 | 63·5 | 73·42 | 75 | 75·75 |

*Fig.* 7.5.

Now that we have these results we can begin to analyse them. We should expect that a shorter step length will give an improvement in the accuracy of the solution, and this is borne out in the table. At what stage does the time taken for the solution become a more important consideration than the degree of accuracy? The second method gives appreciably better results than the first method, even using quite a large step length. Is this going to be so for all equations? In this way the pupil is encouraged to investigate further even if a complete answer is not within his grasp at his present stage of mathematical education.

These are only a couple of examples of the work that can be done; there are many other possibilities. Statistics is another field where the availability of a computer can completely change the approach. A suite of standard statistics programs to plot histograms, calculate means and standard deviations and calculate correlations and regression lines has obvious advantages, not the least being that the pupil can work with convincing statistics. The computer can use its ability to provide random numbers to provide data, to test mathematical models—repeating the same data to test an improved model, and then to provide a series of tests on the improved model.

Finally it must be said that the computer will bring us closer to tackling real problems in the sixth form. It will give the pupil some insight into what mathematics at the industrial level is all about. He may come to have some experience of indeterminacy in a problem and will become more accustomed to making investigations rather than producing neat, and often useless, solutions.

## Bob Lewis  **Mathematics and Science**

Before moving on to the computer solution of problems, let us look at the more fundamental problem of creating a mathematical model describing a scientific situation. 'It is rarely necessary to introduce a mathematical idea without the basis of a real physical situation already appreciated by the student. A mathematical model can be built up which describes this and then can be extended and tried in different situations. Gradually an appreciation that these apparently different physical concepts can be described in a similar way to one another brings home the underlying mathematical structure which is common to them all. One most important point is that the examples used must be credible and that they must extend to as many corners of the physical, biological and social sciences as possible.'*

The definite effort to involve credible situations should provide sufficient motivation for a student interested in science to be prepared to spend some time on mathematics. This approach was used in a monograph 'Growth and Decay Models', introducing first-order differential equations and now in trial form in a number of schools.

Careful consideration of the scientific factors involved leads to the postulation of a simple model. This model can be tested, and the sophistication of the testing can be a lead into tests of significance

* J. Hyslop and R. Lewis, 'Mathematical Models in Science'. *Int. J. Math. Educ. Sci. Technol.* 1970.

and the other factors to be taken into account if the simple model fails. This approach can be outlined as follows:

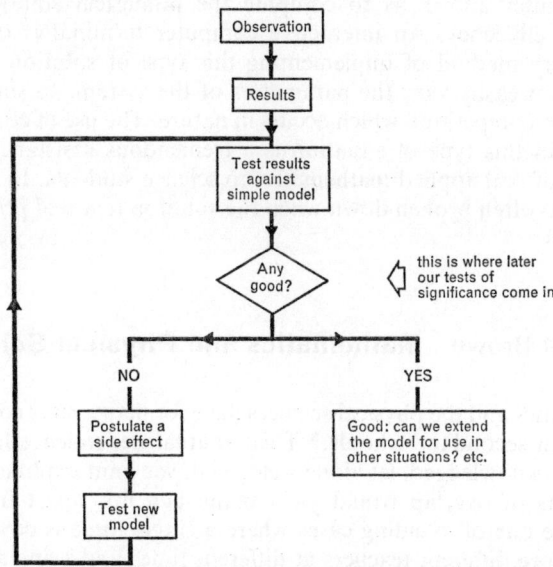

Taking a particular example from biology, the growth of a species dependent on binary fission leads to a very simple model. With limited nutrient the basic equation is

$$\frac{dx}{dt} = kx(a - x)$$

where $x$ is the density of the species

$k$ is a constant related to the rate of binary fission

and $a$ is the maximum density which can be supported by the given nutrient.

The analytic solution of this equation is straightforward for A level students, but if we consider the slightly more complex case of two species competing for the same nutrient we modify our model for species X to allow for the density of species Y and see how much of the nutrient that species will require.

So we get

$$\frac{dx}{dt} = \frac{R_1 x}{K_1}(K_1 - x - Ay)$$

and

$$\frac{dy}{dt} = \frac{R_2 y}{K_2}(K_2 - y - Bx)$$

The analytic solution of these simultaneous differential equations is not simple and so we turn to numerical solutions. It is here that the computer allows us to compute the numerical solution with ease and efficiency. An interactive computer terminal is the most satisfactory method of implementing this type of solution and the student can easily vary the parameters of the system, so simulating the type of competition which occurs in nature. The use of computing facilities in this type of situation is a tremendous assistance in the teaching of real applied mathematics to science students. In the past this has so often broken down when the solution to a real problem is attempted.

## Margaret Brown   **Mathematics and Physical Science**

Mathematics and the physical sciences have for many years co-existed uneasily in secondary schools.* Their mutual dependence has been scarcely acknowledged, let alone welcomed, yet joint exploitation of their areas of overlap would yield many benefits, apart from the immediate one of avoiding cases where a single topic is covered by two or more different teachers at different times and using different notations. Not the least of the benefits would be, for the mathematics teacher, the extra motivation provided by the scientific experiment and, for the science teacher, the advantage of having the mathematics he requires provided at the right moment by a mathematical specialist.

One area which has formed the subject of such a joint mathematics-science module of work is that of molecular size† in which the pupil is introduced naturally to index notation and standard form in his quest to determine the approximate length of a molecule. On the way he finds how many times he can cut a piece of paper in half, and hence the size of the smallest piece of paper he can obtain. This provides an opportunity to introduce the use of the $2^n$ and $2^{-n}$ notation. He then adds 1 grain of potassium permanganate to 1 litre of water, and continues to dilute the solution by a factor of 10 until the colour is no longer visible. This gives a very rough upper limit to the size of a molecule of potassium permanganate, and incidentally introduces the $10^n$ and $10^{-n}$ notations.

The final dénouement involves an oil-drop whose volume is approximately $0.5$ mm$^3$, which momentarily forms a circular patch

* G. Matthews and M. Seed. 'The Co-existence in Schools of Mathematics and Science.' *Int. J. Math. Educ. Sci. Technol.* 1970.
  † Nuffield Mathematics Project module 'Indices and Molecules'.

when allowed to touch a water-surface. By assuming that the oil-molecules behave like a pile of marbles and spread out to form a layer which is one molecule thick, the length of an oil molecule can be estimated. This provides a further opportunity for the use of standard form, which is reinforced by examples involving on the one hand sizes and masses of different molecules, and on the other distances and masses in astronomy. An added bonus is that in order to estimate the length of the oil-molecule the pupil must be able to determine the area of a circle, and this can be an excuse to introduce $\pi$.

One can see that here the mathematics and science are interwoven, and can therefore naturally be developed together. The same is true in many other cases; for example, networks and circuits; lenses, enlargement and similarity. Other topics might include rate of reaction (or cooling), speed and gradient; waves and trigonometry; logarithms, growth and decay; switches and logic; mirrors and symmetry; and so on.

In such cases an interdisciplinary approach serves to illuminate both the scientific subject matter and the mathematical concepts introduced.


## Margaret Brown   Mathematics in Sociology

Mathematics has much to contribute to the study of biological and physical sciences. However, it is also true today that the social sciences, economics, psychology, sociology and so on, are also making increasing use of mathematical ways of thought.

The following example, for instance, shows how mathematical ideas were used to compare racial attitudes in different areas. Instead of using a questionnaire, which would still require statistical analysis, but would not be easy to draw up on such a delicate topic, it was decided to try to find a way of looking at a situation where people of different races are present in order to determine a way of measuring the degree of integration that occurs.*

Suppose we begin with the simple case of a queue of people, such as might occur in a factory canteen, college refectory, or even a row of children in a school playground. The following diagrams show two such queues in which white and shaded spots indicate the different racial origins of the people in the queue. We have taken only two races in order to simplify the situation. The question we

* Campbell, Kruskal and Wallace. *Seating Aggregation as an Index of Attitude.* Sociometry, 1966.

wish to solve is which one shows the higher degree of racial integration, and for this we need to ask what measure we can use to compare
them.

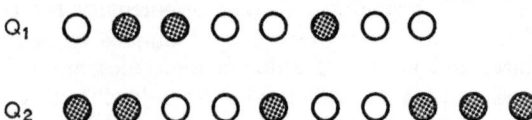

It is clear that our index will depend to some extent on the total
number of people in the queue, which we shall call $N$, and on the
number of each race.

We shall call the number of white spots $W$, so that we have that
for $Q_1$, $N = 8$ and $W = 5$, whereas for $Q_2$, N = 10 and $W = 4$.

To simplify the pattern we will simply consider the nature of the
racial origin of the neighbours of each person in the queue, which
comes down to a consideration of each neighbouring pair. In each
queue of $N$ people there must be $N - 1$ such neighbouring pairs,
each one consisting of any person except the last one in the queue,
and the one behind him. We will call the number of such pairs
which consist of one member from each race $M$, so that we have

| $Q_1: N = 8$ | $Q_2: N = 10$ |
|---|---|
| $W = 5$ | $W = 4$ |
| $M = 4$ | $M = 4$ |

It is difficult to compare the values of $M$ directly because of the
different lengths of the queues and the different proportions of the
two races present, but we can in each case take as our 'origin' the
number of pairs we would expect if the people lined up completely
at random, independently of any racial consideration. This can
conveniently be simulated (either directly or by computer) for any
values of $N$ and $W$ by putting $N$ beads, $W$ of them of one colour
and the remainder of another, into a bag and drawing them out at
random, placing them in a line. If this is repeated sufficiently many
times, an estimate, which we shall call $\bar{M}$, can be calculated of the
mean number of racially mixed pairs in a random queue for that
particular value of $N$ and $W$.

This suggests that we take our index of racial integration as
$M - \bar{M}$, which represents the difference between the actual number
of racially mixed pairs and the number we would expect to occur
under similar but random conditions.

However, this would give the same value for the cases where we
obtained, say, 45 mixed pairs instead of an expected 50 and where
we obtained no such pairs instead of an expected 5. As there is

clearly more racial segregation present in the second situation it is necessary to standardise our values of $M - \bar{M}$ by dividing by a number $\sigma$, which represents the expected variability of $\bar{M}$. In fact $\sigma$ can be obtained in the bead simulation by finding the square root of the mean squared deviation from the mean, $\bar{M}$, of the $M$s obtained in each separate trial. This is technically known as the standard deviation.

We then have

$$I = \frac{M - \bar{M}}{\sigma}$$

as our index, and we can obtain $\bar{M}$ and $\sigma$ experimentally using the differently coloured beads.

However, there is also a way in which we can calculate $\bar{M}$ and $\sigma$. For instance the probability of picking a white bead out of the bag is $W/N$. The probability that out of the $N - 1$ left, the next one will be coloured is

$$\frac{N - W}{N - 1}.$$

Hence the probability of a white followed by a coloured bead is

$$\frac{W}{N} \cdot \frac{N - W}{N - 1}.$$

Similarly the probability of picking a coloured bead followed by a white one is

$$\frac{N - W}{N} \cdot \frac{W}{N - 1}.$$

The probability that any pair is mixed is therefore equal to the sum of these, namely

$$\frac{2W(N - W)}{N(N - 1)}.$$

As there are $N - 1$ pairs, the number of these we would expect to be mixed, $\bar{M}$, is obtained by multiplying $N - 1$ by this expected proportion, giving

$$\bar{M} = \frac{(N - 1)2W(N - W)}{N(N - 1)} = \frac{2W(N - W)}{N}$$

We can, in a similar but more complicated way, also obtain the expected value of the standard deviation in the random case, which gives us

$$\sigma = \sqrt{\frac{\bar{M}(\bar{M} - 1)}{N - 1}}$$

For instance in the case of $Q_1$ and $Q_2$

$$Q_1: \bar{M} = \frac{10 \times 3}{8} = 3.75 \qquad Q_2: \bar{M} = \frac{8 \times 6}{10} = 4.8$$

$$\sigma = \sqrt{\frac{3.75 \times 2.75}{7}} \qquad \qquad \sigma = \sqrt{\frac{4.8 \times 3.8}{9}}$$

$$= 1.21 \qquad \qquad = 1.42$$

$$I = \frac{M - \bar{M}}{\sigma} \qquad \qquad I = \frac{M - \bar{M}}{\sigma}$$

$$= \frac{4 - 3.75}{1.21} \qquad \qquad = \frac{4 - 4.8}{1.42}$$

$$= 0.21 \qquad \qquad = -0.56$$

These two values indicate that the first queue is slightly more integrated than one would expect from a random queue with the same numbers of each race, but that the second queue is less integrated than the equivalent random situation.

By examining the formation of many such queues in each of two or more areas, it is possible to compare their mean index to decide whether there is a significant difference in the degrees of racial integration in those areas. One can build up on similar lines a more complex model which can take account of seating patterns in classrooms, or patterns of house ownership in different streets, and so on.

The same model can be applied, for example, to determine the degree of mixing of the sexes under certain conditions. This is indeed a feature of the power of mathematics, that the same structure can be applied to many apparently diverse situations. It may also be remarked that the above example could be tackled at different levels of sophistication—younger secondary children who would not appreciate the symbolism above could still obtain $\bar{M}$ and $\sigma$ by counting beads.

## Brian Dudley   Mathematics and Biology

Two examples are given to illustrate something of the range of possibilities available in the area where mathematics and biology can overlap.

### 1. *A study of laurel leaves*

Every living thing must have a steady supply of energy if it is to stay alive and this comes ultimately from the energy in sunlight. However, it is only green plants, and especially their leaves, which are able to convert this sunlight energy into a suitable form for use by living things. The questions 'How do leaves grow?' and 'What value are any findings to the biologist?' are two which bring mathematics and biology together.

Since all laurel leaves appear similar, whatever their size, they might be growing by enlargement. This can be confirmed by testing two such leaves and finding a centre of enlargement. What is the biological significance of this result? Since enlargement is the only growth factor involved then there is no cell division and no differentiation at this time. Therefore, one can expect the number of cells in a particular leaf to be constant. All its parts would then have to be formed in the bud, making the number of stomata in each leaf constant, though their number per unit area must fall as the leaf enlarges while their diameters increase. These matters are important when considering the water balance of a leaf and its rate of transpiration. Since there is a maximum size to a leaf, enlargement does not continue indefinitely; the size and the number of cells in it must each have a limit. Is the final size of a leaf determined principally by the size attained by the cells or by their number?

It becomes a simple matter to derive a method for calculating the area of any laurel leaf because its area grows in the same way as for any enlarging shape. This opens up many aspects of plant growth for investigation in the classroom. More mathematics is involved in testing each of the above conclusions.

## 2. *The mathematics of muscle performance*

Large animals have considerably more muscle than small ones have. Does this mean that the former can perform greater muscular feats?

One of the simplest situations that can be used to study both the performance of muscle and the effects of size upon it is that of the jump, or hop, of a locust. At first it is possible to investigate such questions as 'How many times its own length can a locust jump?' If humans could do the same, how far would this be, what would have to be their take-off speed (initial velocity) and therefore their landing speed? How safe would it be for humans if they could perform as locusts do?

Nearer the sixth form level one can develop the facts that the pre-adult locust has no functional wings, behaves as does a projectile once it is airborne and so complies with Newton's laws of motion. From there it is possible to derive equations which will give the initial velocity, the duration of the jump and the highest possible jump of a locust. Other formulae can be used to derive the duration of the contraction of muscle, the kinetic energy involved and the power generated.

By studying locusts at different stages in their life-cycle and also a range of adults of different animals—locust, frog and human, for instance—one can begin to answer such questions as 'What factors, if any, limit the power of muscles?' 'Are the muscles of all animals equally powerful?' 'Is an animal able to perform greater muscular feats as it grows to maturity?'

One approach is to use the mathematics to predict the results and to decide beforehand what data need to be collected. The experiments then confirm or modify in some way the expected result. Another approach is to gather the experimental data first, using them to answer these questions empirically and then seeking mathematical relations for the results afterwards. In either case the pupils (and the teachers) are in for some surprises and some entertainment.

# 8 Summing up: Maths with Everything

Geoffrey Matthews

In the previous seven chapters, we have come a long way, from nursery classes to A level in fact, and it may not be easy to follow an underlying thread of thought common to all the stages concerned. There is, however, a very definite message which may emerge from a quick 're-cap'. From the start of school up to the age of about 11 the children are in the stage of 'concrete operations', that is they learn by *doing*. For infants, 5 to 7, and indeed 'nursery' children, 3 to 5, it is very difficult to isolate mathematics from their other activities. Some practice is, of course, necessary but the mathematical ideas can arise naturally, for example from a post-office project or interest in Guy Fawkes or Christmas activities.

In the junior school, again mathematics is made meaningful by contact with the environment, for example the study of design or architecture. For the middle years, say 9 to 13, the pattern is the same. The children are gradually attaining the stage of 'formal operations' when they can eventually do their problems in their heads and on paper, without the use of 'concrete' materials, but still the problems are whenever possible related to their experiences. The 'early leavers' will learn mathematics only if it is shown to be purposeful, whether linked to their cookery or woodwork or thought in some way to be 'useful' later on in their career.

Its only for the exam candidates that the problems become artificial and the techniques sometimes apparently purposeless, but even here the lesson is being learnt. For example, at A level attempts have been made, notably by Mathematics in Education and Industry, to relate the mathematics with real life rather than the contrived frictionless pulleys and endlessly oscillating particles of the past.

A general thread now becomes apparent: school mathematics must keep one eye on the outside world. The first reforms were carried out within the subject, and rightly so as there was so much to re-think. Idiotic problems about leaking cisterns and ditch-fillers, routines for factorisation, parroted 'theorems' of doubtful validity: these had obscured the nature of the subject and the big ideas had to be unearthed and updated. But the next stage is to start looking across the boundaries, and the last chapter shows some starts that have been made.

We have had a glimpse of the way ahead: 'maths with everything'. But in our enthusiasm we must not submerge mathematics totally within the general curriculum. There are still topics which are straight *mathematics* and there are still times when honest practice is necessary. But if mathematics and biology, economics . . . are to retain their individual fortresses let us now explore the highways between them—that is the message for the next few years.

And while we are dreaming of computers and inter-disciplinary collaboration at advanced levels, let us not forget that mathematics is with us from the beginning. If it starts with the mumbo jumbo of table-chanting and 'two minuses make a plus', many children will never recover from this early abuse: if, on the other hand, it grows naturally from the environment the children will also grow to enjoy the subject and to thrive at it.

We are back finally where we should be: talking about *children*. For mathematics is literally useless without people to create it, to use it, to enjoy it; and 'Mathematics through School' means that as we go forward re-thinking the curriculum we must constantly bear in mind that the new mathematics is for new children.

# Notes

The main secondary projects are described in the booklet *Mathematics Projects in British Secondary Schools*, G. Bell & Sons Ltd for the Mathematical Association, 1968. The 'Schools Council Secondary School Mathematics Project' mentioned there has since been re-named 'Mathematics for the Majority', whose materials are being published for the Schools Council by Chatto and Windus. There is also a 'Mathematics for the Majority Continuation Project', directed by Peter Kaner from 3 The Cloisters, Cathedral Close, Exeter.

The teachers' guides and other materials of the Nuffield Mathematics Project (age range 5 to 13) are published jointly by John Murray and W. & R. Chambers; details are available from Chambers at 11 Thistle Street, Edinburgh EH2 1DG.

The Nuffield Project has also made these short films (15 to 20 minutes) suitable for showing at Colleges, parent/teacher meetings, etc.

| Title | Purchase from | Hire from |
|---|---|---|
| *Maths With Everything* (Infants) | Graphic Films Ltd<br>1 Soho Square<br>London, W.1 | Concord Films Council<br>Nacton<br>Ipswich, Suffolk |
| *I do and I understand* (Juniors) | Sound Services Ltd<br>Wilton Crescent<br>Merton Park<br>London, S.W.19 | Petroleum Film Bureau<br>4 Brook Street<br>London, W.1 |
| *Into Secondary School* (Secondary) | ditto | ditto |

Professional associations which teachers and students may wish to join:

The Mathematical Association, 150 Friar Street, Reading RG1 1HE

Association of Teachers of Mathematics, Market-Street Chambers, Nelson, Lancashire

Institute of Mathematics and its Applications, Maitland House, Warrior Street, Southend-on-Sea, Essex SS1 2JY